LADY MARGARET HALL LIBRARY
OXFORD

D1348475

305139222R

YALE UNIVERSITY
MRS. HEPSA ELY SILLIMAN
MEMORIAL LECTURES

Germ Cells and Soma: A New Look at an Old Problem

ANNE McLAREN

NEW HAVEN AND LONDON
YALE UNIVERSITY PRESS

Table 1 originally appeared in M. H. L. Snow, *Journal of Embryology and experimental Morphology 42* (1977):292–303, and figure 12 originally appeared in A. J. Copp, *Journal of Embryology and experimental Morphology 51* (1979):118; both are used with the permission of the authors and the *Journal*. Table 3 first appeared in M. H. L. Snow and P. P. L. Tam, *Nature 279* (1979):555–57, and figures 13 and 22 in M. Monk and M. I. Harper, *Nature 281* (1979):311–13; all are used with the permission of the authors and *Nature*, copyright © 1979 Macmillan Journals Limited. Table 4 is adapted from R. L. Gardner, in *Birth Defects* (1977):154–66, and is used with the permission of the author and Excerpta Medica. Figure 3 originally appeared in C. D. Turner, *General Endocrinology* (1966):400, 5th ed., and is reproduced with the permission of W. B. Saunders. Figure 5 is adapted from R. M. Moor and D. G. Cran, in *Development in mammals* (1980):3–37, volume 4, and is used here with the permission of the authors and Elsevier-North Holland Inc. Figure 7 is reproduced from R. M. Moor and G. M. Warnes, in *Control of ovulation* (1978), with the permission of the authors. Figure 10 is reproduced from T. Ducibella, T. Ukena, M. Karnovsky, and E. Anderson, *Journal of Cell Biology 74* (1977):156–57, with permission from the authors and the *Journal*. Figure 15 is reproduced from S. J. Kelly, *Journal of experimental Zoology 200* (1977):367, with permission from the author and the *Journal*. Figure 19 is adapted from P. Tam and M. H. L. Snow, "Proliferation and migration of primordial germ cells during compensatory growth in the mouse embryo," and is used with the permission of the authors.

Designed by Sally Harris
and set in Times Roman type.
Printed in the United States of America by
The Alpine Press Inc., Stoughton, Mass.

Library of Congress Cataloging in Publication Data

McLaren, Anne.
Germ cells and soma.

(Mrs. Hepsa Ely Silliman memorial lectures ; 45)
Based on 3 lectures delivered by the author at Yale University in 1980.
Bibliography: p.
Includes index.
1.Developmental cytology—Addresses, essays, lectures. 2. Cell differentiation—Addresses, essays, lectures. 3. Cellular control mechanisms—Addresses, essays, lectures. 4. Germ cells—Addresses, essays, lectures. 5. Somatic cells—Addresses, essays, lectures.
I. Title. II. Series. [DNLM: 1. Germ cells. WQ 205 M478g]
QL963.5.M35 591.3′3 81–2971
ISBN 0–300–02694–3 AACR2
10 9 8 7 6 5 4 3 2 1

THE SILLIMAN FOUNDATION LECTURES

On the foundation established in memory of Mrs. Hepsa Ely Silliman, the President and Fellows of Yale University present an annual course of lectures designed to illustrate the presence and providence of God as manifested in the natural and moral world. It was the belief of the testator that any orderly presentation of the facts of nature or history contributed to this end more effectively than dogmatic or polemical theology, which should therefore be excluded from the scope of the lectures. The subjects are selected rather from the domains of natural science and history, giving special prominence to astronomy, chemistry, geology, and anatomy. The present work constitutes the forty-fifth volume published on this foundation.

This book is dedicated to all who generously
gave me permission to quote their unpublished
work and particularly to my colleagues in the
MRC Mammalian Development Unit, Paul Burgoyne,
Marilyn Monk, Mike Snow, and David Whittingham,
whose work provided the inspiration for the
lectures on which this book is based.

Contents

1
The Problem

The problem of germ cells and soma is one that is deeply rooted in nineteenth-century biology—but it is none the worse for that. Why return to it at this time? As I hope to show, enough new findings on the development of the germ cells and the early embryo have accumulated in recent years to make it worth looking again at the problem, with particular reference to mammals.

Two possible relationships between the germ cells and the soma, or body, are set out in figure 1. The upper diagram is usually associated with the name of August Weismann, who coined the phrase "the continuity of the germ plasm" (Weismann 1892). Earlier, Nussbaum had postulated that there was an actual continuity between germ cells of succeeding generations, a continuous germ-cell lineage. He wrote: "The reproductive cells of the higher animals represent the stock from which the individuals after a short existence detach themselves to die, like the leaves falling from a tree" (Nussbaum 1880). When this theory was disproved, Weismann put forward the less precise, less easily testable concept of germ plasm. Ac-

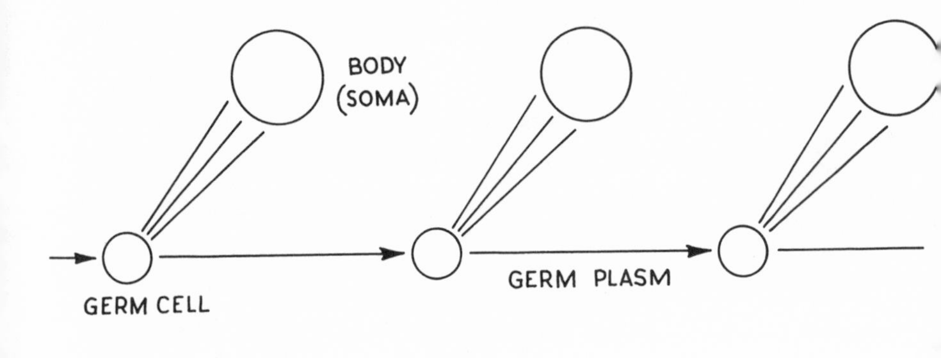

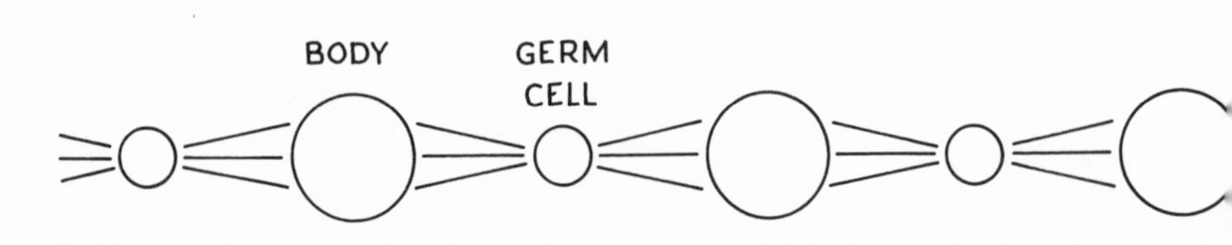

Figure 1. Two Contrasting Views of the Relation between Germ Cells and Soma. The upper diagram reflects the view of August Weismann, the lower that of Charles Darwin. (Adapted from Michie 1958, figure 6.)

cording to this hypothesis, a continuous stream of immortal germ plasm, concentrated at intervals in actual germ cells, buds off a mortal soma in each generation. Influences on the germ plasm can affect the soma, but there is not—and in principle there cannot be—any reciprocal influence of the soma on the germ plasm.

In striking contrast is the relationship sketched in the lower diagram. We are taught in school that the function of the germ cells is to reproduce the species—in other words, that an egg is a chicken's way of making another chicken. Samuel Butler turned this thought upside down and asserted that a chicken was an egg's way of making another egg. With this diagram, both statements have equal validity, since the life process is seen as cyclical. Influences on the germ cells can still affect the soma, but now the soma can exert a reciprocal influence on the germ cells. This was the scheme that Charles Darwin believed in, and indeed it was in this context that he put forward his ingenious theory of pangenesis, according to which germ cells are formed by self-replicating "gemmules" aggregating together from all tissues of the body. This theory constituted one of the earliest postulations of self-replicating particles in the history of genetics, and although today it sounds fanciful, it is typical of Darwin that he should have made the first serious attempt to integrate embryonic development and genetics, cell heredity and germ-line heredity.

The implications of these two schemes for genetics, and their translation into the terms of nucleus and cytoplasm, or nucleic acid and protein, have been considered by others (see, for example, Michie 1958; Darlington 1953). My aim is rather to summarize what we now know about the development of germ cells and soma. Figure 2 corresponds more closely to developmental reality than do either of the conventional diagrams; in its philosophical implications it is closer to Darwin's scheme. The epiblast, as we shall see in chapter 4, is the business part of an embryo, after it has formed its extraembryonic membranes but before it has started to form any organs. The dotted arrows in the diagram represent the hypothetical influence of the soma on the developing germ cells; I shall mention as we come to them possible routes by which such influence might be exerted.

If one is talking about the development of the mouse, it is conventional to begin at fertilization and continue through to the adult. If one is talking about the development of the egg, it is conventional to begin with the primordial germ cell and end at ovulation. Since we are concerned with the entire process as a continuous cycle, we could begin anywhere. I have decided to disregard the two conventional starting places and

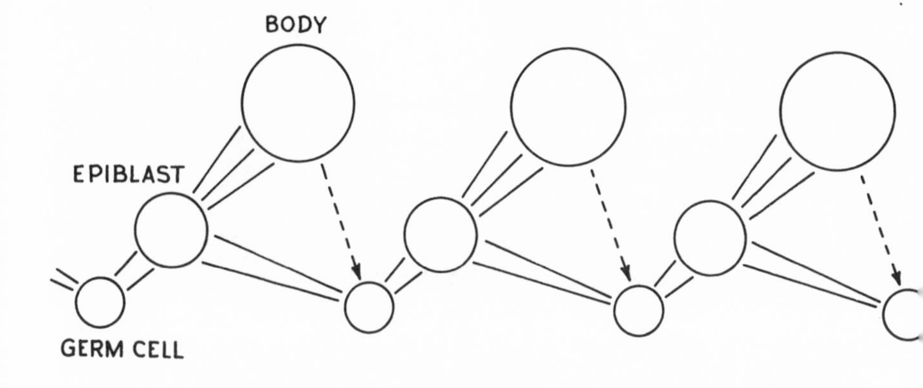

Figure 2. An Alternative View of the Relation between Germ Cells and Soma. The epiblast is the precursor of the fetus, at the pregastrulation stage. The dotted arrow represents the sum total of all the influences that the somatic tissues exert on the developing germ cells.

to begin our story at the moment when the female germ cell first becomes ensconced in its follicle in the ovary and the follicle starts to enlarge. In chapter 2, I shall consider the so-called resting stage of the oocyte and its growth, maturation, and ovulation. Chapter 3 will deal with the activation or fertilization of the egg. Chapters 4 and 5 will cover the development of the preimplantation embryo, the postimplantation development up to gastrulation, and the emergence of the primordial germ cells. Chapter 6 will follow the germ cells on their journey to the future gonads, and chapter 7 will discuss the factors that determine the fate of a germ cell within the gonad up to the stage of the resting oocyte in its follicle within the ovary, which is where we began. Along the way, I shall highlight what seem to me to be the most interesting themes, not attempting to give a comprehensive account of any aspect of development; I shall also point out the gaps in our knowledge, the chief unsolved mysteries. In the last chapter, I shall put forward my own conjectures, with the aim of stimulating research designed to prove me wrong.

2
How the Mouse Makes the Egg

The female germ cell embodies a most profound contradiction. It must be totipotent, containing within itself all the information needed for the developmenι of a new individual. But it must also be highly differentiated in order to carry out the complex functions required of it. We first meet it partway through this differentiation process.

Within the ovary, the oocytes exist in very intimate association with somatic cells—that is, with the follicle cells that surround them. In the mouse, the follicle cells can already be seen arranging themselves around the oocytes on the day before birth. By 3–4 days after birth, the oocytes are arrested with their chromosomes in meiotic prophase. A single layer of flattened follicle cells surrounds each oocyte to form a primordial follicle. Tens of thousands of primordial follicles are formed in each infant ovary, and though many degenerate before puberty, several thousand persist into adult life.

Figure 3 shows a diagrammatic cross section of an adult

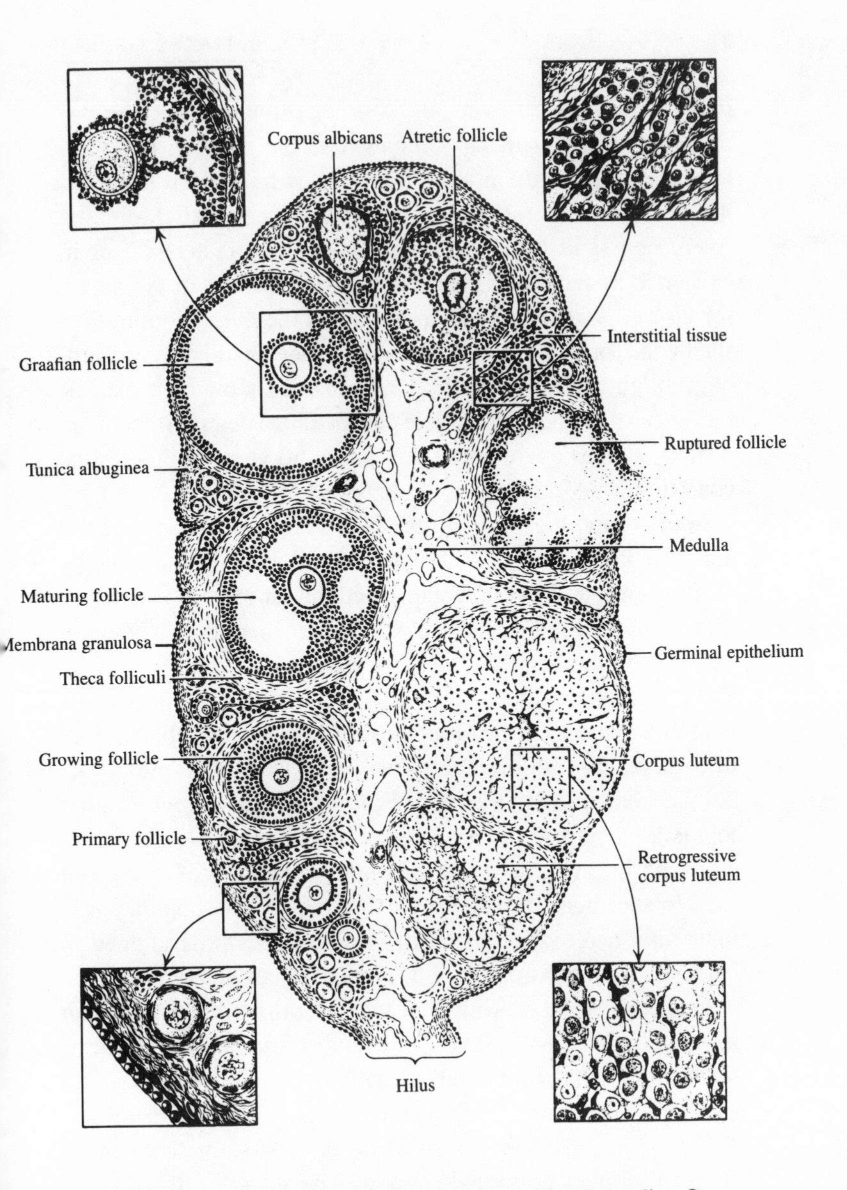

Figure 3. The Development of Follicles in the Adult Mammalian Ovary. After follicular rupture, the *corpus luteum* is formed and then regresses. (From Turner 1966, figure 12–7.)

ovary. The most obvious and striking structures are the medium and large growing follicles, and the *corpora lutea* that are formed from them once the oocytes have been shed; but the most numerous structures are the small primordial follicles. Every egg shed in adult life has spent the greater part of its existence in such a follicle; oocytes may persist in the mouse for up to 2 years, since in most strains the ovary continues to shed eggs for a considerable time after the ageing of the uterus has brought female reproduction to an end. In our own species, the stock of primordial follicles is not exhausted until the menopause, so an oocyte may spend up to 50 years in this resting condition.

In the primordial follicle, the plasma membranes of oocytes and follicle cells are closely apposed and may be connected by desmosomes and even by gap junctions, at least in the mouse (Anderson and Albertini 1976). Whether there is any intercellular communication at this stage we do not know. Nor do we know what determines when a follicle leaves the primordial pool and starts to grow. Statistically, it seems that the number of follicles leaving the pool at any time is proportional to the number remaining in the pool (Faddy et al. 1976), but whether this is a stochastic process, like radioactive decay, so that in any time interval there is a constant probability of any given follicle starting to grow, or whether the follicle population is inherently heterogeneous, with some follicles programmed to leave the pool before others, we again do not know.

In any case, once a follicle leaves the primordial pool, it can never go back. The follicle first changes from a single layer of flattened cells to a cuboidal epithelium, then to a multilayered structure. The large multilayered preantral follicle, the so-called stage 5 follicle, seems to be the last stage that can be attained without hormonal support. The stage 5 follicle is for the first time capable of responding to gonadotrophins, in particular to follicle-stimulating hormone (FSH) produced by the pituitary. If FSH is absent, all the stage 5 follicles degenerate; if it is present, some continue developing to form mature, hollowed-out Graafian follicles. The number of Graafian follicles that are formed depends on the amount of FSH available.

The recent work of Baker, Challoner, and Burgoyne (1980) has established that it is the size of the primordial pool in the mouse as a whole that determines the number of growing follicles, not the size of the pool in an individual ovary. In a normal female, the ratio of the population of growing follicles to the pool size increases as the female gets older; that is, the ratio increases as the pool size decreases (Krarup, Pedersen, and Faber 1969). If this were determined within the individual ovary, removing the other ovary should make no difference. But in fact, if one ovary is removed, thus halving the primordial pool size, the ratio of the population of growing follicles to the pool size in the remaining ovary soon becomes significantly greater than it would be in each ovary of an intact mouse of the same age. It corresponds to the ratio that would normally be seen in an older female, in which the total pool size for both ovaries has dwindled to that remaining in the younger female after unilateral ovariectomy. So the value of this ratio must depend on the total pool size, not just the pool size in one ovary.

So much for the follicle; how about the oocyte inside it? In the first 2 weeks of follicle growth, the oocyte increases more than 300-fold in volume; then it stops growing, and during the later stages of follicle development, no further increase in oocyte size takes place. The early stages of oocyte growth do not seem to require a normal intact follicle. In sex-reversed mice, one can sometimes find growing oocytes within the testis tubules during the first couple of weeks after birth (McLaren 1980); although the oocyte is the same size as it would have been in an ovary of the same age, the surrounding follicle, if one can call it that, is very poorly developed and consists of just a few scattered cells sticking to the oocyte surface (figure 4). A similar situation is seen when infant mouse ovaries are cultured in the absence of gonadotrophins; the oocytes grow, but without any proper follicular development (Baker and Neal 1973). If gonadotrophins are added to the culture medium, the follicles develop normally. Perhaps in both these situations—the testis tubule and the gonadotrophin-deficient culture—the oocyte obtains some nutrients by diffusion from the surrounding somatic tissue. Medium-sized (50μ) oocytes degenerate when isolated in culture, even when cocultured with granulosa cells; but if the intact follicles are cultured, the oocytes grow at the normal rate (Eppig 1978). So at least the later stages of oocyte growth require normal oocyte-follicle contacts.

What are these normal oocyte-follicle contacts? In the normal ovary, as the oocyte grows, the zona pellucida is laid down between the surface of the oocyte and the surrounding follicle cells. Labelling studies suggest that the zona material is secreted mainly, if not entirely, by the oocyte (Haddad and Nagai 1977). The follicle cells send out numerous processes, which penetrate the zona pellucida and terminate on the oocyte membrane, often in bulbous swellings that make deep indentations in the oocyte surface. Specialized junctions at the point of contact between the oocyte and the follicle cell processes are common, and in several species, gap junctions have been reported. It is these cellular processes and gap junctions that are thought to be responsible for the nutritive role that the follicle cells play in oocyte growth. Through the gap junctions, the germ cells and the adjacent follicle cells are electrically coupled; they can also be shown to be metabolically coupled. Fluorescein injected into rat oocytes gradually spreads into the adjacent follicle cells (Gilula, Epstein, and Beers 1978). If sheep oocytes in intact follicles are allowed to attach *in vitro* to a

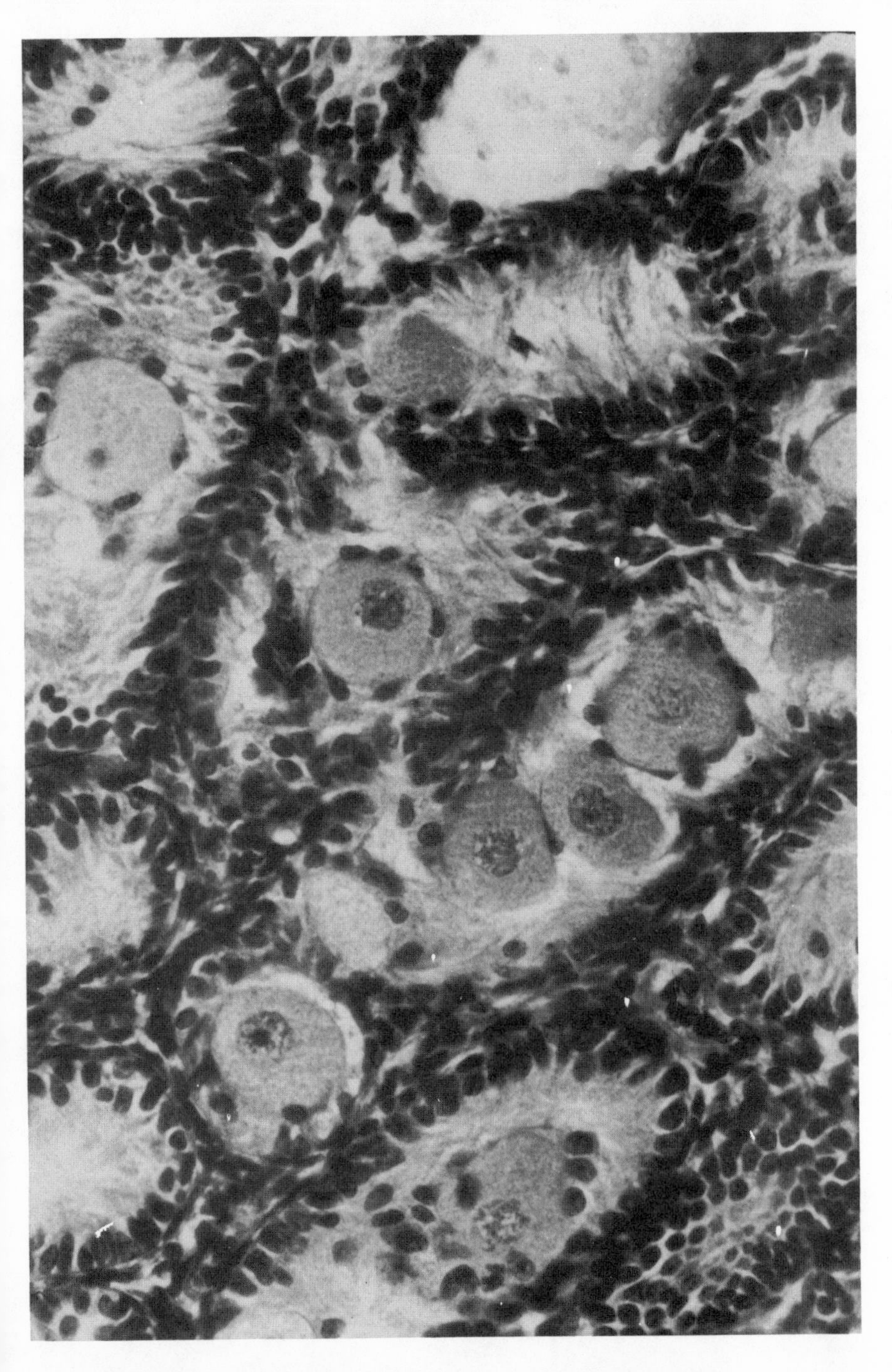

Figure 4. Growing Oocytes in the Testis Tubules of a *Sex-reversed* XX Male Mouse 11 Days after Birth. The "follicles" are very poorly developed and consist of no more than a few scattered cells attached to the oocyte surface.

monolayer of follicle cells prelabelled with radioactive choline, the choline will pass from the monolayer into the follicle cells surrounding the oocyte and then into the oocyte cytoplasm, even though choline is excluded from direct entry into the oocyte because it cannot cross the oocyte membrane (Moor, Smith, and Dawson 1980). Other substances used to demonstrate intercellular coupling between cumulus cells and oocytes include [^{3}H]uridine and inositol. Figure 5 shows a diagram of the oocyte-follicle cell complex and indicates the sites at which hormones, for example, could interrupt metabolic coupling. We shall see presently that this is important in oocyte maturation.

What about macromolecules? There is an upper limit to the size of molecules that can cross gap junctions, but larger molecules could be transferred from the follicle cell processes into the oocyte by pinocytosis. In insects, there is good evidence for the passage of RNA, as well as substances of low molecular weight, from the nurse cells into the developing oocytes. This has not yet been shown in mammals, but there is circumstantial evidence that polypeptides are taken in by the oocyte during its growth phase. Measurements of protein synthesis in the growing mouse oocyte have recently been reported by two groups: Canipari, Pietrolucci, and Mangia (1979) determined the rate of leucine incorporation into total oocyte protein in oocytes exposed to different concentrations of radioactive leucine so as to expand the intracellular leucine pool, while Schultz, Letourneau, and Wassarman (1979) measured the rate of incorporation of radioactive methionine into total oocyte protein and the size of the endogenous methionine pool. Although the two sets of data were obtained in different ways, they agree remarkably well. Canipari and her colleagues concluded that protein synthesis increases linearly with cell volume, but they did not look at oocytes smaller than 40μ in diameter. The data for the growth period as a whole (figure 6)

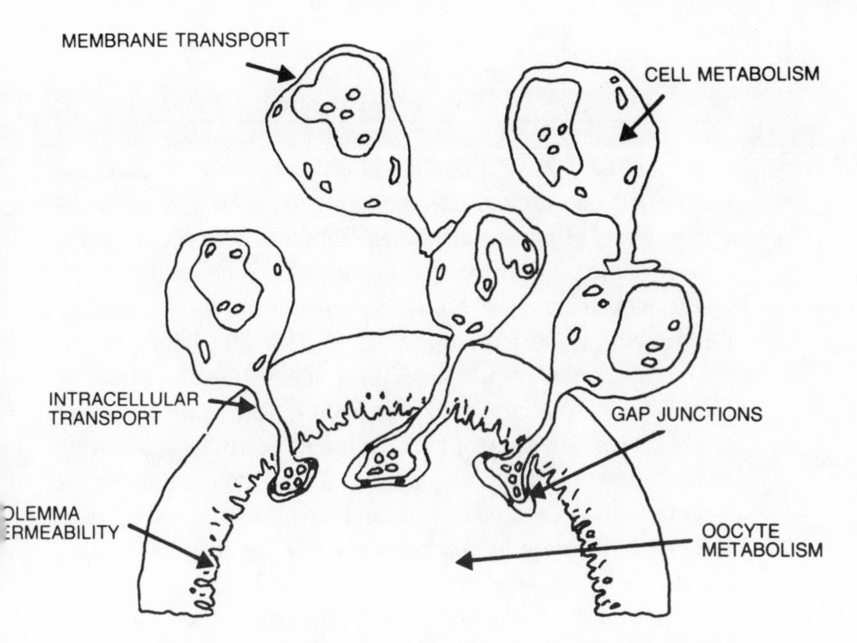

Figure 5. Diagram of the Cumulus-Oocyte Complex Showing the Various Intracellular Sites at Which Hormones and Other Agents Might Disrupt Intercellular Transmission of Metabolic Products. (From Moor & Cran 1980, figure 5.)

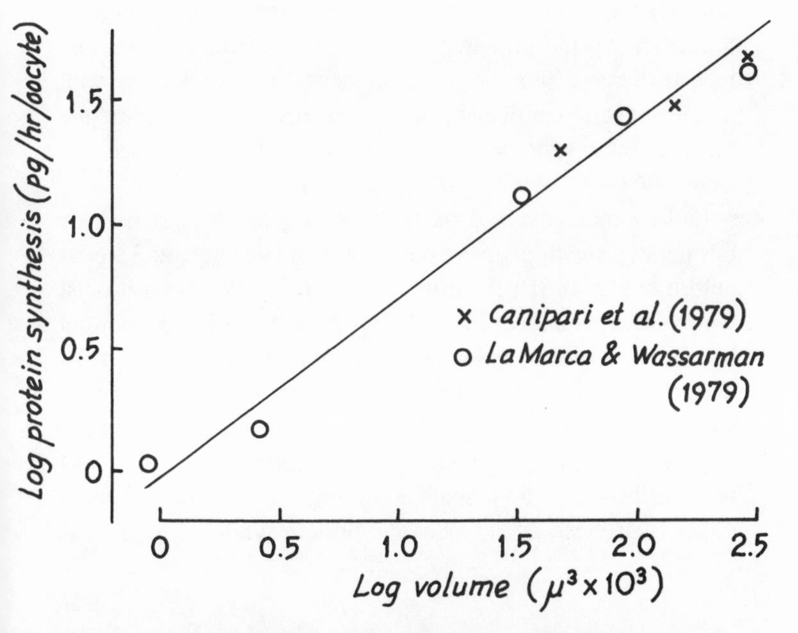

Figure 6. Rate of Protein Synthesis during Growth of Oocytes in the Mouse. The data, taken from LaMarca and Wassarman (1979) and Canipari, Pietrolucci, and Mangia (1979), are plotted logarithmically.

suggest either two straight lines of different slopes, one for the earlier and one for the later growth period, or a steadily decreasing rate of protein synthesis. With the second hypothesis, a logarithmic plot gives a reasonable linear relationship.

The total increase in oocyte volume during the 2 weeks of the growth period is of the order of 350-fold, while the increase in protein synthesis is only about 35-fold. Thus as LaMarca and Wassarman (1979) point out, the rate of protein synthesis per unit of cytoplasm decreases nearly 10-fold during oocyte growth. They calculate that in the course of these 2 weeks, each oocyte can synthesize about 6.6 ng of protein, and that in the ensuing week, up to ovulation, the fully grown oocyte can synthesize another 6.2 ng, making a total of some 12.8 ng of protein. But at ovulation the egg contains about 25 ng of protein. In other words, in the time available, the oocyte's own protein-synthetic machinery is only capable of making about half the total protein that the mature oocyte contains (Schultz, Letourneau, and Wassarman 1979). The other 50 percent of the protein must be taken in from outside, and indeed there is immunological evidence that foreign, as well as native, serum antigens can be transferred from the maternal blood to the oocyte within the follicle (Glass 1961; Glass and Cons 1968).

In the mouse, the bulk of the RNA in ovulated eggs is made during the growth phase of the oocyte, between 1 and 3 weeks before ovulation (Bachvarova and De Leon 1980), so it must be unusually stable. Little RNA persists from the primordial follicle, and little is made in the mature oocyte. At the end of the growth phase, the germinal vesicle of the oocyte contains numerous so-called "nuclear bodies," which are made up of large amounts of accumulated ribonucleoprotein. Large numbers of ribosomes are assembled during oocyte growth, but the majority of these are not active in protein synthesis in the mature oocyte. The ovulated egg contains about 500 pg of total RNA, including some 25 pg of messenger RNA—more than enough to program all the egg ribosomes.

Oocyte maturation

Before the egg is ovulated, it has to undergo maturation. FSH not only stimulates the final burst of follicular growth and the formation of the mature Graafian follicle, it also induces the formation of luteinizing hormone (LH) receptors in follicle cells. In response to LH, the oocyte resumes meiosis, the chromosomes move from meiotic prophase into the first meiotic division, germinal vesicle breakdown occurs, and the first polar body is extruded. By the time of ovulation, the egg chromosomes are once again in a state of meiotic arrest, this time in the second meiotic metaphase. The follicle cells also respond to LH: in particular, the follicle cell processes that pass through the zona pellucida are interrupted, so that the oocyte and the follicle cells are no longer electrically or metabolically coupled. Estrogen is replaced by progesterone as the dominant follicular steroid, the follicle cells luteinize, and the egg is shed.

The interruption of intercellular coupling must exert a profound effect on oocyte function. It has been suggested that the follicle cell processes and the gap junctions through which they communicate with the oocyte hold the chromosomes in meiotic arrest by transmitting an inhibitory substance. The most likely candidate for the inhibitor is cyclic adenosine monophosphate (AMP), since isolated rat and mouse oocytes can be prevented from resuming meiosis by exposure to cyclic AMP (for references, see Dekel and Beers 1978). There are problems with the hypothesis that the disruption of the gap junctions is directly responsible for the resumption of meiosis, however, since in response to LH release, meiosis seems to be resumed at a stage when the junctional complexes are still morphologically intact (Szöllösi et al. 1978) and still capable of transferring choline from the follicle cells to the oocyte (Moor, Polge, and Willadsen 1980). A more likely hypothesis would be that the follicle cell processes control cytoplasmic maturation in the oocyte, since the disappearance of the gap junctions is followed immediately by a reduction in charge on the oocyte surface and migration of the cortical granules to their definitive position beneath the cell membrane. The temporal relationship between interruption of the gap junctions and peripheral migration of cortical granules holds true not only for mouse oocytes but also for cow, pig, and rabbit oocytes (Szöllösi et al. 1978).

When the follicle cell processes are interrupted artificially, by physically removing the oocytes from their follicles and isolating them *in vitro*, all these maturation processes follow in their normal sequence. Indeed, a small proportion of mouse oocytes matured *in vitro* can develop successfully to late gestation or even birth after *in vitro* fertilization (Cross and Brinster 1970; Mukherjee 1972). The capacity of the oocyte itself to resume meiosis seems to a large extent unrelated to follicular development, since even oocytes from preantral follicles will resume meiosis when isolated in culture (see Thibault 1977), provided the oocytes themselves are above a certain size (Iwamatsu and Yanagimachi 1975). The conditions for successful *in vitro* maturation are still incompletely known, however. When isolated rabbit, pig, or cow oocytes are matured in culture, spermatozoa can penetrate them, but the oocyte cytoplasm does not transform the sperm head into a normal male pronucleus. It seems that some cytoplasmic factor—termed by Thibault (1972) the male pronucleus growth factor (MPGF)—fails to be synthesized if the gap junctions between the oocyte and follicle cells are interrupted before, or too soon after, the LH surge. Addition of hormones to the culture medium leads to an increase in sperm head swelling.

RNA synthesis ceases shortly before germinal vesicle breakdown (Rodman and Bachvarova 1976), but polypeptide synthesis continues throughout maturation. The pattern of polypeptide synthesis shows both quantitative and qualitative changes, so that each stage of maturation is associated with the appearance of specific polypeptides, some of which are synthesized for very brief periods only (Van Blerkom and McGaughey 1978). The inhibition of germinal vesicle breakdown prevents these changes, suggesting that the synthesis of polypeptides involved in maturation may be triggered by the mixing of cytoplasm and nucleoplasm. The pattern of synthesis appears to be affected by the hormonal milieu in which maturation takes place. This is particularly important, because the proteins made during this period seem to play a key role, not only in maturation, but also in subsequent development. Sheep follicles removed from the ovary and matured *in vitro* show alterations to the normal profile of steroids that they secrete. These alterations lead to the synthesis of a different pattern of polypeptides in the oocytes and subsequently to anomalies of fertilization, delay in cleavage, and failure of blastocyst formation (Moor 1978; Moor, Polge, and Willadsen 1980). The addition of gonadotrophins alone to the culture medium has little effect, but with estrogen too, normal development to the blastocyst stage is possible (figure 7). Indeed, lambs have now been born from the fertilization of oocytes matured *in vitro* in the presence of gonadotrophins and estrogen (Moor and Trounson 1977).

The changes in polypeptide synthesis that occur in isolated sheep oocytes *in vitro* are not identical to those seen *in vivo* or *in vitro* in intact follicles treated with estrogen (Warnes, Moor, and Johnson 1977). The isolated oocytes lack the proteins nor-

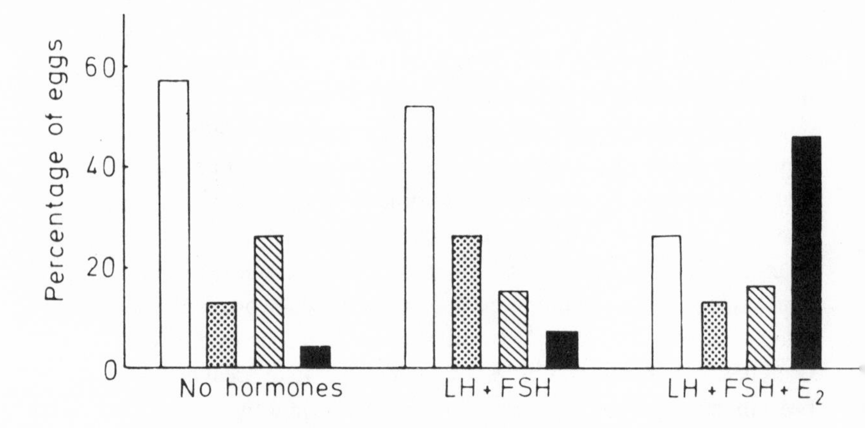

Figure 7. Effect of Gonadotrophins (LH, 1 μg/ml; FSH, 2 μg/ml) and Estrogen (Estradiol-17β, 1 μg/ml) on the Subsequent Development of Sheep Oocytes Cultured in Follicles for 24 Hours before Transfer to Inseminated Recipients. One week after transfer, the embryos were classified in the four categories indicated below. (From Moor and Warnes 1978, figure 10.10; after Moor and Trounson 1977, table 4.)

- ☐ = single-celled
- (stippled) = fragmented
- (hatched) = cleaved but retarded
- ■ = fully differentiated normal blastocysts

mally synthesized in the later stages of maturation. These proteins are presumably involved less with nuclear maturation than with the cytoplasmic changes needed for successful fertilization and embryonic development. Although oocytes matured *in vitro* are morphologically indistinguishable from those matured *in vivo*, it is becoming increasingly clear that their developmental capacity is likely to be seriously impaired.

Somatic influence

The time span from the formation of the primordial follicle soon after birth to the maturation and ovulation of the egg that it contains covers a period of about 40–400 days in the mouse,

10–50 years in man. What is the evidence that, during this period, the female leaves her mark on the eggs that she is making? Harking back to figure 2, do any "dotted lines" of somatic influence upon the developing germ cells exist? If they did, how could we detect them?

First, there is the possibility that as the mother ages, the oocytes deteriorate, because of the extended period that they have to spend in first meiotic prophase. In a classical experiment on the small invertebrates known as wheel animalcules, or rotifers, Lansing (1952) bred two lines, one propagated entirely from eggs laid by elderly females, the other propagated entirely from adolescent mothers. In each generation, the "o'd" line developed more quickly and died at a younger age, until finally the line died out, suggesting that the effect of senescence on the egg cytoplasm is transmitted from one generation to the next. As far as I know, this experiment has never been repeated in mice. It has been claimed in regard to mammals that eggs from older mothers are of poor quality, but most of the claims have not been confirmed. For example, Parkening and Chang (1976) reported that the eggs of aged female hamsters took an abnormally long time to be fertilized, but further investigation showed this to be an effect of senescence on the maternal environment rather than on the eggs themselves.

One possible instance of oocyte deterioration involves the increased incidence of chromosomal trisomy in embryos from older mothers. This has been found in both mouse and man, though only certain chromosomes show the increase. In Man the effect of maternal age proves to be most pronounced for trisomies involving the small chromosomes (both acrocentric and nonacrocentric). An exception is chromosome 16, which shows a very high rate of nondisjunction, but little or no maternal age effect (Hassold et al. 1980). The classic example of a maternal age effect is of course trisomy for chromosome 21,

Down's syndrome, where the chances that a woman will have an affected baby, either one that survives to birth or one that dies early in pregnancy, more than doubles every 5 years when the mother is over the age of 30. The father's age is now thought to be irrelevant (Erickson 1978). Some of the suggested explanations do not involve deterioration of the eggs with age, but most of these have either been disproved or have failed to be substantiated. For example, the suggestion by German (1968) that reduced frequency of intercourse might result in delayed fertilization in older women, and hence cause increased nondisjunction, is not compatible with the observed irrelevance of the father's age, nor with the finding that most of the relevant nondisjunction occurs at the first meiotic division, well before fertilization. The "production line" theory of Henderson and Edwards (1968), which postulates that the oocyte population is heterogeneous and that those oocytes ovulated late in adult life are those that have entered meiosis late in fetal life, have fewer chiasmata, and hence are more liable to nondisjunction, has so far received little experimental confirmation.

It is also possible that the maternal age effect not only has nothing to do with oocyte ageing but even has nothing to do with oogenesis. Recent reports (see Erickson 1978) suggest that in a substantial proportion of trisomies for chromosome 21, the extra chromosome comes from the father. If the proportion of paternally to maternally derived trisomies proves to be independent of maternal age—and this information is not yet available—then the explanation of the maternal age effect must be sought after fertilization. Perhaps the more efficient uterus of a younger woman rejects trisomic embryos before the pregnancy is even recognized; in the older uterus, such embryos may more often be aborted later or even survive until birth, when they manifest Down's syndrome.

Time will show whether any of these explanations turns out to be true. At present it seems more likely that we are dealing with one of the dotted lines on our diagram—in other words, that defects in the oocyte, induced by the somatic environment, progressively accumulate during the long period that the oocytes spend in meiotic prophase, making nondisjunction more likely to occur. For example, Ford (1960) suggested that the normal breakdown of the nucleolus toward the end of the first meiotic prophase might be delayed by some ageing factor; the nucleolus might then interfere with chromosome pairing and increase the incidence of nondisjunction.

Apart from this possible increase in susceptibility to nondisjunction, the oocyte seems well protected during its long sojourn in the primordial follicle from any effects of toxic chemicals in the maternal environment or hormonal disturbances, for example; but as ovulation approaches, it becomes much more vulnerable. Efforts to find new contraceptives have led to an increased understanding of the various ways in which ovulation can be prevented and the oocytes diverted away from the pathway of maturation into the pathway of atresia and degeneration. Fortunately, these are usually all-or-nothing effects; but we are also beginning to see how disturbances in the hormonal environment during oocyte maturation may cause abnormalities in development after fertilization—as in the work of Moor and his colleagues on sheep oocytes (Moor, Polge, and Willadsen 1980). So the oocyte may not possess so great an immunity to environmental influences as has sometimes been suggested.

Cytoplasmic effects

So much for the changing environment that the oocyte endures. What about the effects of the genetic constitution of the mother? What influence does this exert on the cytoplasm of the egg? What about the 50 percent of cytoplasmic proteins that appear to be introduced into the oocyte from outside? What proteins are they? Are they concerned only with maturation and ovulation, or are they important for development?

If they are important for development, and if they are subject to genetic variation, they would presumably be manifest as cytoplasmic effects. Such effects could usefully be studied in experimental mouse chimeras, where an oocyte of one genotype may be surrounded and fed by follicle cells of a contrasting genotype. Would the oocyte thus nourished show any modification of cytoplasmic type?

It turns out that very few genetic effects exerted through the cytoplasm have so far been detected in mammals (for review, see McLaren 1979). Some maternal effects on gene penetrance have been detected but not yet analyzed, involving, for example, various agouti locus genes (Wolff 1978) and the X-linked gene *Tabby* (Kindred 1961); these effects could turn out to be cytoplasmic, but in the case of such late-acting genes it seems unlikely. An apparent cytoplasmic effect, in which the viability of embryos heterozygous for the gene *Hairpintail* ($T^{Hp}/+$) depends on which parent transmits the mutant gene, is open to an alternative interpretation, in terms of inactivation of the paternally derived chromosome region (see McLaren 1979). Strain differences in the response of mice to the teratogen 6-aminonicotinamide seem to be cytoplasmically inherited (Verrusio, Pollard, and Fraser 1968), but the effect here may be associated with a genetically determined difference in mitochondria.

One case that seems worth pursuing from this point of view is the case of the Japanese mouse strain DDK (see review by McLaren 1976a). DDK females, although fully fertile with males of their own strain, show very high embryonic mortality if outcrossed. Wakasugi (1973, 1974) has evidence to suggest that a cytoplasmic factor is manufactured by the oocyte or is taken in from the follicle cells. Chimera studies might shed light on the source of the cytoplasmic factor. Would a DDK oocyte matured in a follicle made up largely of non-DDK cells still show the characteristic DDK incompatibility on outcrossing?

The fact that so few genetic effects exerted through the cytoplasm have so far been identified does not necessarily mean that the oocyte cytoplasm has no effect on development. As we shall see in chapter 4, it is the very early stages of development—the preimplantation stages—that are most likely to be affected. We know that preformed messenger RNA in the egg cytoplasm directs development during early cleavage, and that abnormal females—for example, XO females with only one X chromosome, instead of the two that are normally both functional during oogenesis—may produce abnormal eggs and hence abnormal embryos. We also know that certain strains of mice show more preimplantation mortality than others and that embryos of some strains can be successfully cultured from the 1-cell stage, others only from the 2-cell stage. These more general effects on growth and viability may well turn out to be influenced by cytoplasmic factors, including perhaps some factors that are passed into the egg from the follicle cells.

We now possess the techniques to investigate this possibility, so in the next few years we may achieve an increased understanding of the somatic contribution to germ cell development. Even at the present time, an extreme Weismannist view cannot be upheld. The mouse does make the egg; and the properties of the mouse and the circumstances of the mouse can affect the developmental potential of that egg.

3
Activation and Fertilization

The maturation of the oocyte is complete once its chromosomes are arrested in the metaphase stage of the second meiotic division. This arrest can be ended only by an activating stimulus. Although activation is normally triggered by the fertilizing spermatozoon, many other effective activating stimuli are known: for example, electric shock, heat shock, exposure to calcium ionophore or reduced concentrations of calcium, injection of calcium into the egg, mechanical stimuli, exposure to enzymes, alcohol, anesthetics, tranquillizers, and protein synthesis inhibitors (for reviews, see Graham 1974; Whittingham 1980).

Activation is characterized by two simultaneous events: the resumption of meiosis, leading to extrusion of the second polar body, and the release of the cortical granules into the perivitelline space. Both events seem closely associated with changes in the concentration of free, as opposed to bound, calcium. How can the very varied stimuli known to produce activation

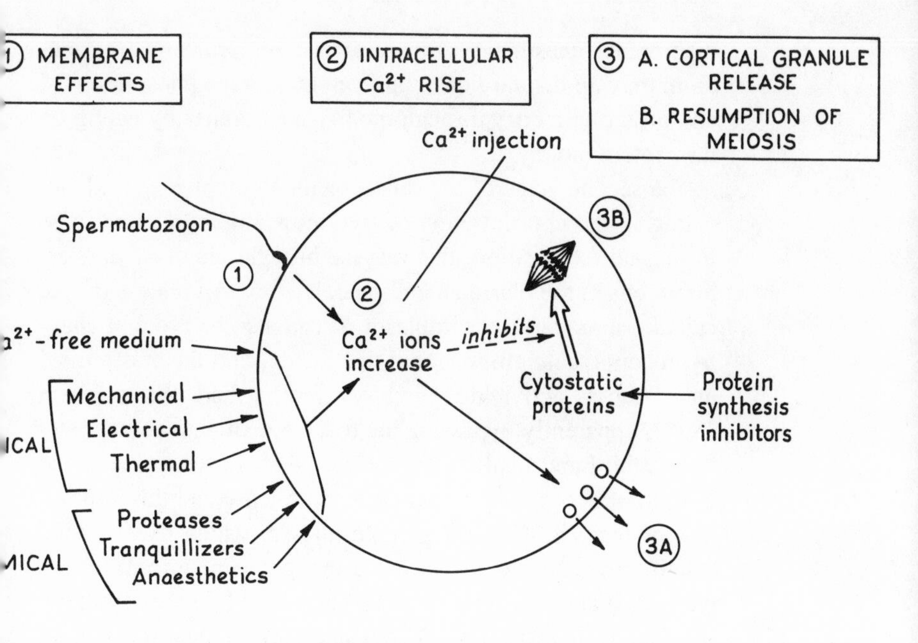

Figure 8. Oocyte Activation: The Agents That Can Induce It and the Stages at Which They May Act. (After Whittingham 1980, figure 5).

bring about these two events? Much research is being devoted to activation at present, and it seems that different stimuli act at different levels (see figure 8).

The primary activating stimulus must involve some perturbation of the plasma membrane of the oocyte, either producing a transient depolarization of the membrane—as has been shown to occur in sea-urchin eggs at fertilization (Epel 1978)—or perhaps displacing calcium from the membrane (Whittingham and Siracusa 1978), or both. The fertilizing spermatozoon perturbs the oocyte membrane in such a way as to produce fusion between its own plasma membrane and the oocyte membrane. Other activating stimuli may also perturb the membrane in some way: electric shock can cause depolarization; thermal shock can affect the lipid component of the membrane; anesthetics, tranquillizers, and calcium-free medium may all displace calcium from membrane phospholipids; and proteolytic enzymes can modify membranes by acting on the protein moiety.

The second stage of activation seems to involve a rapid elevation in the concentration of free intracellular calcium ions. In normal fertilization, this may be brought about by the perturbations of the plasma membrane, leading to a release of free calcium ions into the cytoplasm. It can also be brought about by injecting calcium ions into the oocyte, and this has proved an effective activating stimulus (Fulton and Whittingham 1978), apparently bypassing the first stage of activation.

The third and final stage involves what we recognize as the criteria of activation: cortical granule release and the resumption of meiosis. Cortical granule release, which plays a vital role in preventing polyspermy, is thought to be a form of calcium-mediated exocytosis, and so may result directly from the increase in free calcium. The resumption of meiosis is more complicated. In Amphibia, the arrest of the oocyte chromosomes in second meiotic metaphase has been shown to be due to a cytostatic factor, protein in nature, which is destabilized at high calcium concentrations (Meyerhof and Masui 1977). So the increase in free calcium ions in the oocyte may inhibit the factors that are responsible for the block to meiosis. The activating influence of protein synthesis inhibitors could operate merely by stopping the production of these same cytostatic factors. We do not yet know whether protein synthesis inhibitors have any effect on cortical granule release.

Parthenogenesis

When the egg is activated in ways other than by a spermatozoon, subsequent embryonic development is termed parthenogenetic. In one strain of mice (LT) the eggs activate spontaneously, either after ovulation or even while still in the ovary. In the latter case, activation occurs after the oocyte has completed the first meiotic division, and parthenogenetic development within the ovarian follicle proceeds fairly normally up to an early egg cylinder stage. Embryonic organization then breaks down, and an ovarian tumor, or teratocarcinoma, is formed (Stevens and Varnum 1974). With ovulated eggs, activated either spontaneously or artificially, the parthenogenetic embryo will be diploid if the expulsion of the second polar body is prevented, or if the chromosome set is doubled after activation; otherwise it will be haploid. In either case, development may proceed apparently normally—though perhaps rather slowly—up to, and even beyond, implantation.

There seems to be no specific point in development at which parthenogenetic embryos fail, no specific developmental process that they cannot undergo; yet no parthenogenetic embryo has developed to term, and the oldest to have been reported was a mid-somite embryo with limb buds and a beating heart (Kaufman, Barton, and Surani 1977). What goes wrong? The individual cells are viable, since aggregation chimeras made between normal and parthenogenetic embryos show survival of parthenogenetic tissue (Surani, Barton, and Kaufman 1977; Stevens, Varnum, and Eicher 1977). Indeed, a female chimera of this type has successfully bred from her parthenogenetic component (Stevens 1978), proving that fertilization is not a necessary condition for the subsequent formation of germ cells, and that the oocyte itself is potentially totipotent in the sense that it can provide all that is required for the development of any cell type in the female body, even more egg cells. Yet the parthenogenetic embryo on its own cannot even develop to term.

The problem was further exacerbated when the successful production of homozygous diploid mice was reported. If either the male or female pronucleus is surgically removed from a fertilized mouse egg (Modlinski 1975) and the remaining chromosomes are doubled up by use of the drug cytochalasin B, so as to restore diploidy, the resulting embryo will develop not merely to the blastocyst stage (Markert and Petters 1977) but to full term, giving rise to a normal fertile mouse with chromosomes derived from but a single parent, which may be either the original mother or the original father (Hoppe and Illmensee 1977). These homozygous diploid mice are always female, since an embryo formed by diploidization of a male pronucleus containing a Y chromosome would contain no X chromosome in its genome, and such embryos, lacking all X chromosome products, do not survive.

When the male pronucleus is removed and the female pronucleus is diploidized, the nuclear status of the embryo is indistinguishable from that of a haploid parthenogenetic embryo that has been diploidized (figure 9). Yet the surgically produced homozygous diploid survives, while the parthenogenote dies. The nuclear adequacy of the parthenogenote is further emphasized by the recent birth of mice following transfer of nuclei (Illmensee and Hoppe 1981) from parthenogenetic blastocysts into enucleated fertilized eggs (Hoppe and Illmensee, personal communication). Perhaps the spermatozoon brings in some factor essential for subsequent development, which remains when the pronucleus is removed. The search for this hypothetical factor is now in full swing in several research groups.

Fertilization

It is primarily the fortunes of the female germ cell that we are concerned with, but having now introduced the spermatozoon, we must look briefly at the developmental history of the male germ cell also.

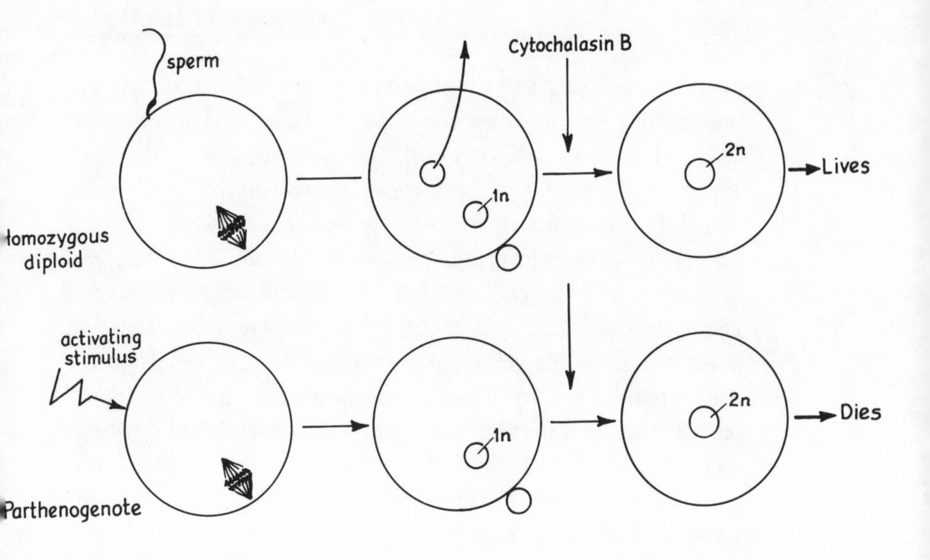

Figure 9. A Comparison between the Mode of Production and Nuclear Status of the Homozygous Diploid and the Parthenogenote. Why does one live and the other die?

Unlike the population of female germ cells, the stem-cell population of male germ cells continues to proliferate throughout life. Successive spermatogenic waves leave the stem-cell population and multiply by mitosis; they then enter meiosis and undergo maturation through spermatocyte and spermatid stages, finally transforming from spermatids into spermatozoa, with the loss of almost all the original cytoplasm. Within the testis, germ cells and somatic cells are very intimately associated, with spermatids almost entirely embedded in the giant Sertoli cells. Such a situation would seem to provide ample opportunity for somatic influence on the course of spermatogenesis. However, studies on mouse chimeras, where germ cells and somatic cells can be of different genetic constitution, have so far failed to provide convincing evidence of any somatic influence on sperm genotype (McLaren 1975a) or sperm shape (Burgoyne 1975). On the other hand, chimera studies involving the mutant gene *testicular feminization* have established that the somatic cells need to be sensitive to androgen if normal spermatozoa are to be formed (Lyon, Glenister, and Lamoreux 1975). The spermatozoa themselves can be of a genotype that is totally irresponsive to androgen, provided that at least some of the cells making up the testicular environment can respond normally. Thus the somatic environment is important, at least in a permissive sense, for normal sperm development. Elsewhere I have discussed other ways in which male germ cells may be affected by the chemical or biological environment in which they develop (McLaren 1979).

The fertilization process itself has been intensively studied but is still not completely understood. The acrosomal enzymes associated with the sperm head may be involved more in loosening the cumulus mass surrounding the egg than in actually penetrating through the zona pellucida (Bedford and Cross 1978), despite earlier beliefs. The spermatozoon seems to be taken into the egg cytoplasm by a combination of membrane fusion, in which the equatorial segment of the sperm membrane fuses with the plasma membrane of the egg, and subsequent engulfment of the spermatozoon by the egg microvilli (Noda and Yanagimachi 1976). At least in the rabbit, an immature oocyte that has not yet undergone germinal vesicle breakdown does not interact normally with spermatozoa (Berrios and Bedford 1979); the oolemma is receptive, and membrane fusion takes place, but cortical granules are not extruded, the oocyte cortex fails to engulf the spermatozoon fully, the nuclear envelope persists, and the sperm chromatin is not dispersed, since the male pronuclear growth factor is lacking.

If the egg is mature, however, the interaction of the fertilizing spermatozoon with the egg membrane causes activation. As with other activating stimuli, the cortical granules are extruded, producing a zona and/or vitelline reaction that prevents any further spermatozoa from entering the egg. Meiosis is resumed, the second polar body is extruded, and male and female pronuclei become visible within a few hours of the first sperm attachment. The patch of sperm membrane fused into the egg membrane is said to be detectable, using fluorescent labelling techniques, at least up to the 8-cell stage in the mouse embryo (Gabel, Eddy, and Shapiro 1979).

Sperm penetration very rarely occurs in the neighborhood of the meiotic spindle. The egg membrane in this area, encompassing about 20 percent of the cell surface, is devoid of microvilli, and the cortex lacks cortical granules (Nicosia, Wolf, and Inoue 1977).

Although mouse oocytes are capable of synthesizing DNA during maturation, to repair damage to chromosomal DNA caused for example by ultraviolet irradiation (Masui and Pedersen 1975), the S-phase that takes place in the male and female pronuclei after fertilization represents the first scheduled DNA synthesis to occur in the egg since the last premeiotic S-phase before birth. Because the completion of the second meiotic division and the expulsion of the second polar body do not take place until after the spermatozoon has entered the egg cytoplasm, the mammalian egg—unlike the haploid spermatozoon—never has less than a 2c (diploid) DNA content. Nor does the egg ever entirely lose genetic heterozygosity, since segregation at any locus that happens to be separated from the centromere by an odd number of chiasmata does not take place until the second meiotic division, after sperm entry.

4

How the Egg Makes the Mouse

In this chapter, I shall cover that portion of the life cycle during which no germ cells are manifest. It begins with the fertilized egg, itself the fusion product of two germ cells, and ends with the newly gastrulated primitive-streak stage embryo and the emergence of a new generation of germ cells. The end result may not yet bear much resemblance to an adult mouse, but nonetheless the basic architecture of the animal has been laid down: there is a head and a tail, a right and a left side, and three germ layers. As we shall see, the embryonic genome has already been switched on. The egg has done its part, and the embryo's own genes are now controlling its development.

The onset of cleavage

At the end of the last chapter, we left the fertilized egg to undergo its first cleavage division. This fertilized egg contains about 25 ng of protein, of which approximately half has been synthesized within the oocyte, and half has been acquired from the surrounding follicle cells in the ovary. It also contains about 500 pg of RNA, of which some 25 pg is message. This is more than enough messenger RNA to program all the ribosomes in the egg. Only about 25 percent of these ribosomes are at present organized into polysomes, and protein synthesis is at a low ebb.

What evidence is there of spatial heterogeneity within such an egg? There is, of course, a cortical cytoskeleton, made up of microtubules, located beneath the plasma membrane, while the two pronuclei lie in the more fluid cytoplasm towards the center. As we saw at the end of the last chapter, the spherical symmetry of the egg is perturbed in at least two ways. First, before fertilization, the region of the egg membrane overlying the meiotic spindle—about 20 percent of the total surface—is differentiated from the rest by an almost complete absence of microvilli, and the subcortically located granules are also lacking (Nicosia, Wolf, and Inoue 1977). Sperm penetration rarely if ever takes place in this region, and most of the bare area is lost from the surface when the second polar body is extruded (Eager, Johnson, and Thurley 1976). A prominent mid-body persists for some time between the polar body and the vitellus, marking the point of extrusion. Second, sperm antigens can be detected after fertilization by the use of fluorescent antibody techniques and are reported to persist as a discrete patch at least until the 8-cell stage (Gabel, Eddy, and Shapiro 1979). Whether either of these asymmetries is related in any consistent way to the plane of first cleavage or to any other aspect of subsequent development is not known.

Cleavage up to about the 32-cell stage retains a considerable degree of synchrony, but even at the 2-cell stage, synchrony is not perfect. An interval of about an hour on average separates the two cell divisions that transform a 2-cell into a 4-cell embryo. Dissociation and labelling experiments have shown that the descendants of the cell that divides earlier tend to divide before the descendants of the cell that divides later all the way through cleavage, right up to the blastocyst stage; but though at each cleavage division they divide on average earlier than the other cells, they do not divide any more quickly (Kelly, Mulnard, and Graham 1978), so that the average interval of about an hour does not increase. When an 8-cell stage divides to form a 16-cell embryo, one or two of the daughter cells get pushed into the interior of the ball of cells; these tend preferentially to be the progeny of the first cells to divide, which for geometrical reasons are usually in contact with a larger number of cells (Graham and Deussen 1978). This means that in normal cleavage, the first cell to divide from the 2-cell stage tends to contribute a disproportionately larger number of descendants to the inner cell mass of the blastocyst and fewer to the outer trophectoderm.

Compaction

Up to the 8-cell stage, the individual blastomeres tend to be rather loosely associated and to be spherical in shape for most of each cleavage stage. At the 8-cell stage a striking and significant change occurs (figure 10). Each blastomere flattens

Figure 10. Scanning Electron Micrographs of Mid-Cleavage Mouse Embryos before, during, and after the Process of Compaction. (a) Four-cell embryo with one dividing blastomere. The arrow indicates the cleavage furrow. The surfaces of the blastomeres are uniformly covered with microvilli (× 1,500). (b) Compacting 8-cell embryo. An apical localization of microvilli is present on four blastomeres (× 1,500). (c) Three blastomeres of a compacting embryo. Microvilli are seen at the apices (*left side* and *lower right corner*) of two blastomeres (× 3,200). (d) Embryo nearing completion of compaction. Only small areas of smooth membrane remain exposed (× 1,600). (e) Compacted embryo in which no smooth surface is visible (× 2,200). (f) Compacted embryo showing smooth areas between the apical localizations of microvilli and the regions of cell-cell contact (× 1,500). (From Ducibella, Ukena, Karnovsky, and Anderson 1977, figures 1, 3, 4, 6, 7, 8.)

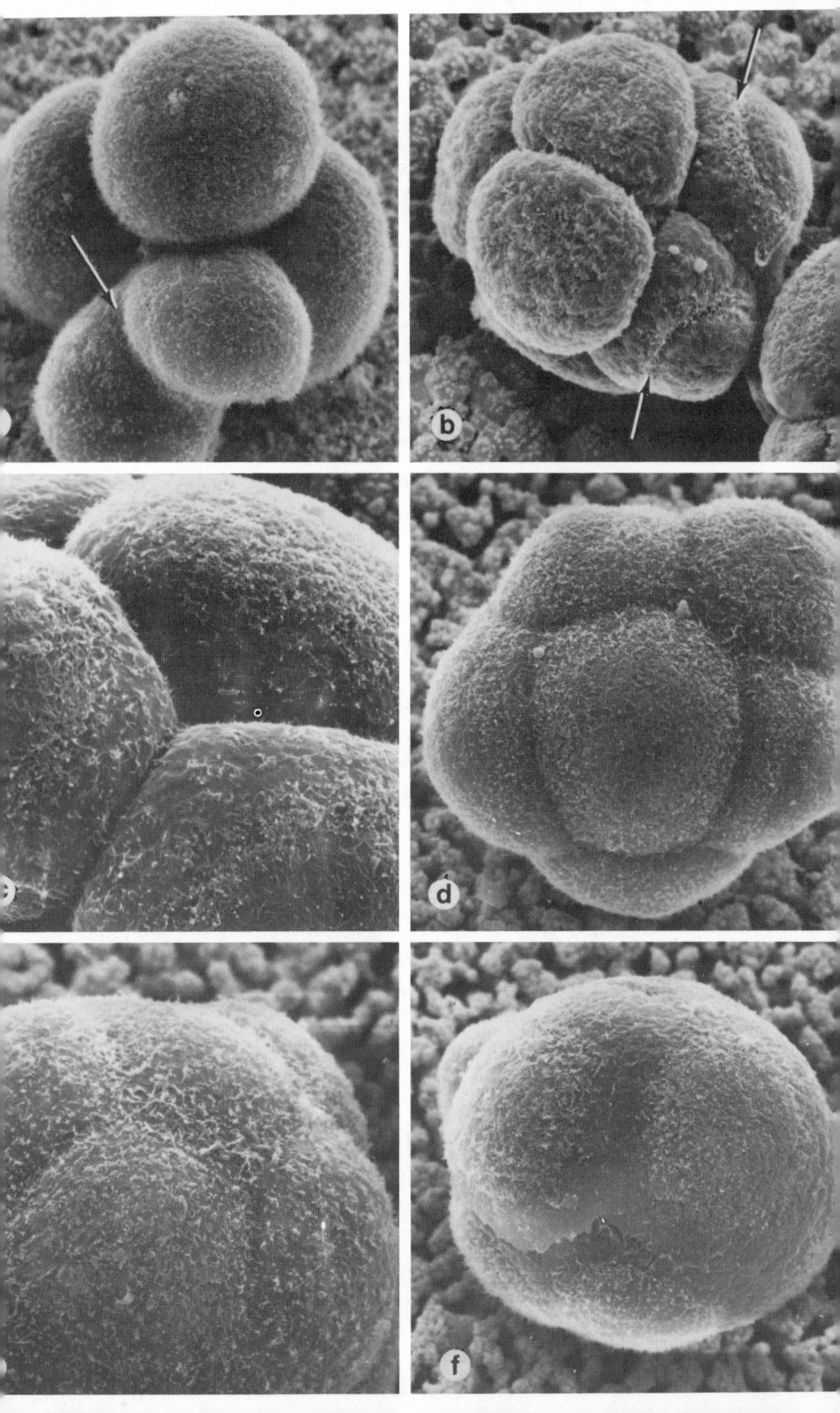
b
d
f

against its lateral neighbors in such a way as to maximize the amount of cell surface contact. This process, termed compaction, alters the shape of the individual blastomeres from spherical to wedge-shaped and for the first time gives them a polarity, so that each cell has an outer and an inner end. At the outer end, focal tight junctions form, linking each cell to its neighbor. The appearance of the embryo changes markedly: the outlines of individual blastomeres can no longer be distinguished, and the embryo as a whole takes on a more or less smooth, spherical form. The distribution of surface microvilli also changes. Within each cell the mitochondria, which at this time are transforming from the vacuolated type characteristic of the oocyte to the more conventional type with many cristae characteristic of most somatic cells, become localized in the cortex, adjacent to the newly established intercellular contacts (Ducibella et al. 1977). Arrays of microtubules can also be seen, aligned parallel to the apposed surfaces of adjacent blastomeres, and it has been suggested that they may be responsible for determining both the location of the mitochondria and the change of cell shape involved in compaction.

During the next two cleavage divisions, the apical junctions develop into junctional complexes, each consisting of a peripheral tight junction, then a more centrally located gap junction, and an occasional desmosome (Ducibella et al. 1975). The gap junctions, which are first established at the 8-cell stage, provide intercellular continuity, as judged by both electrical coupling and dye penetration studies (Lo and Gilula 1980). The cytoplasmic vesicles that are a characteristic feature of the early cleavage stages (Calarco and Brown 1969) coalesce to form fewer, larger vesicles, which then disappear, presumably releasing their contents into the intercellular spaces as the blastocyst cavity forms. The outer layer of cells of the blastocyst—the trophectoderm—pumps fluid inward, and, since the tight junctions form an effective permeability seal—the *zonula occludens*—the blastocyst expands. The mouse trophectoderm at this stage constitutes the earliest example of a true epithelium, with an active transport system that selectively moves chlorine and potassium ions into the fluid-filled blastocoel during its expansion (Biggers and Borland 1976).

Inner cell mass versus trophectoderm

The inner cell mass of the blastocyst forms a striking contrast to the outer trophectoderm layer, not only in morphology but also in function and biochemical makeup. The analysis of these two tissues—the first two differentiated tissues of the embryo—has been made easier by the fact that each can now be prepared in pure form, isolated from the other. The trophectoderm cells, which give rise to the trophoblast and form the fetal part of the placenta, are flattened cells, tightly joined by junctional complexes, capable of pumping fluid and inducing the decidual cell reaction in a suitably primed uterus. The inner cell mass, which gives rise to the fetus as well as to the yolk sac and amnion, consists of rounded cells, less tightly linked to one another but capable of aggregating if dissociated.

The two tissues turn out to be differentiated not only morphologically but also biochemically. The pattern of polypeptides synthesized at different stages of preimplantation development has been analyzed in considerable detail by several groups, using labelled amino-acid precursors and both one- and two-dimensional gel systems. The biggest qualitative shift in polypeptide profile occurs between the 2- and 8-cell stages (Van Blerkom and Brockway 1975), when protein synthesis is still at a very low level. Between the 8-cell and blastocyst stage, the rate of protein synthesis increases several fold. Studies using the transcription inhibitor α-amanitin suggest that a transcriptional event crucial for blastocyst formation takes place between the 16- and 32-cell stages (Braude 1979). When the inner cell mass is analyzed separately from the trophectoderm, most of the polypeptides synthesized, but not all, are common to both tissues: a few inner-cell-mass polypeptides are not found in the trophectoderm, and similarly, a few are found only in the trophectoderm (Handyside and Johnson 1978; Johnson 1979). If the inner cell mass is isolated and allowed to develop further *in vitro*, additional changes in the pattern of polypeptide synthesis can be demonstrated (Howe, Gmür, and Solter 1980).

The normal derivatives of the inner cell mass and the trophectoderm are shown in figure 11. Polar trophectoderm, overlying the inner cell mass, divides more rapidly than mural trophectoderm, so that cells shift from polar to mural regions during blastocyst development (figure 12) (Copp 1978). About 4 days after fertilization, the blastocyst starts to implant in the uterus: the mural trophectoderm, beginning at the point farthest away from the inner cell mass, stops dividing altogether, and the cells undergo repeated cycles of endoreduplication to form the giant cells of the primary trophoblast. These giant cells, which eventually range from tetraploid up to a thousand times the normal DNA content, possess invasive properties and form the implantation interface between the embryo and the uterus. Meanwhile the polar trophectoderm continues to proliferate, probably as a result of an inductive influence from the adjacent inner cell mass, the earliest example of induction in mammalian development. As this group of cells enlarges, the line of least resistance is first of all downward into the blastocyst cavity, to form the extraembryonic portion of the egg cylinder, and then, when the egg cylinder fills the blastocyst cavity, upward between the two closely apposed surfaces of the uterine epithelium, to form the ectoplacental cone (fig-

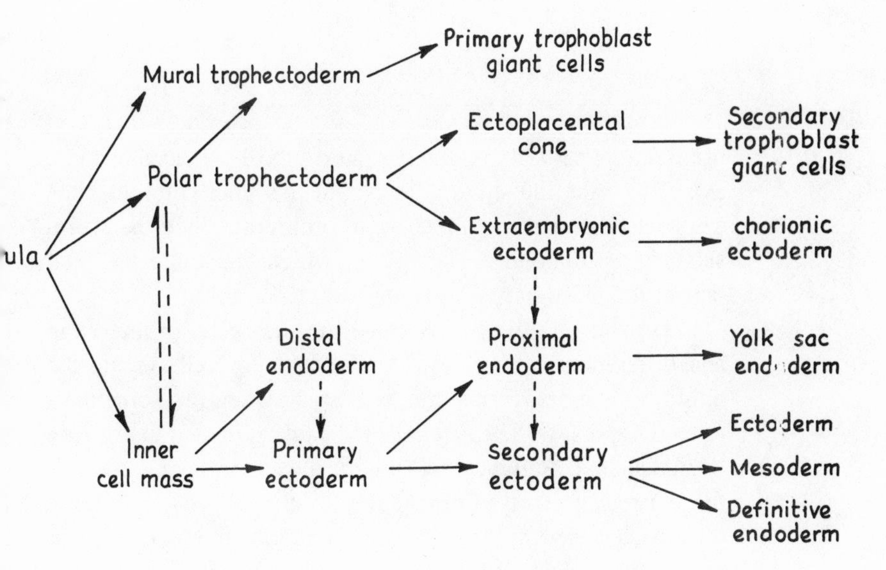

Figure 11. Map of Cell Lineage for the First Week of Development in the Mouse.

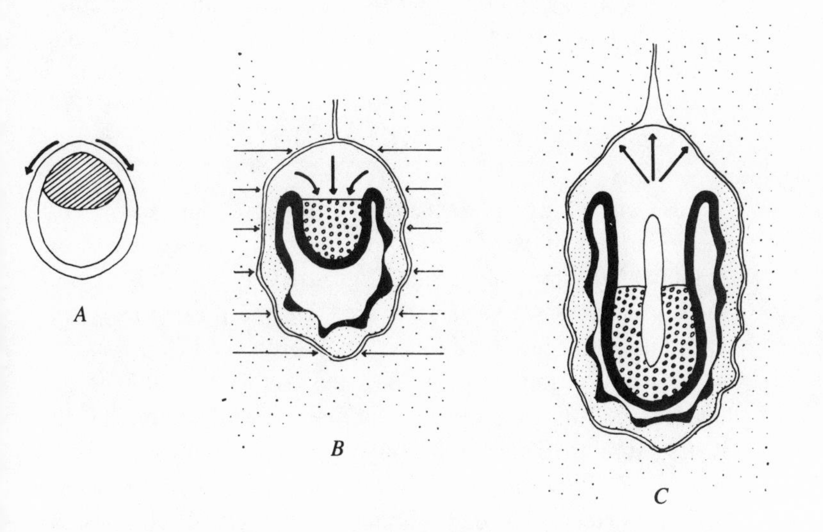

Figure 12. Diagram Showing the Forces Acting during the Development of the Mouse Blastocyst into an Egg Cylinder Stage Embryo. (From Copp 1979, figure 8.)

ure 12) (Copp 1979). At the margins of the ectoplacental cone, away from the inductive influence of the inner cell mass, the trophectoderm cells again cease to proliferate; they then form secondary trophoblast giant cells, which constitute the fetal part of the placenta.

Meanwhile, the inner cell mass too has been proliferating and differentiating (figure 11). The layer of cells facing the blastocyst cavity forms the primary endoderm, sometimes termed hypoblast. The first cells to be differentiated in this way (Dziadek 1979) migrate away to line the entire inner surface of the blastocyst cavity, forming the distal (parietal) endoderm; their successors form the proximal (visceral) endoderm, which covers the rapidly enlarging egg cylinder. This proximal endoderm performs a vital nutritive role for the early embryo. It gives rise to the yolk-sac endoderm, but it almost certainly does not give rise to any of the definitive endoderm of the fetus.

Epiblast

Last, and most important from our point of view, the rest of the inner cell mass goes to form the embryonic ectoderm or epiblast. Why most important? Because it is from this tissue, the epiblast, as we shall soon see, that the primordial germ cells emerge. After the differentiation of the endoderm, about 4–4½ days after fertilization, the epiblast continues to undergo cell division with a doubling time of around 11–12 hours, similar to that found in the preimplantation embryo. About 5½ days after fertilization, the amniotic cavity opens up in the center of the egg cylinder, and the rate of cell division quickens (table 1). Between 6 and 6½ days, the cell cycle time is down to 9 hours; between 6½ and 7 days, to an average of 4½ hours, which is very short for a mammalian cell (Snow

Table 1. Mean Cell Cycle Times Required to Account for the Growth of the Epiblast

	Days *post coitum*								
Age	5½		6		6½		7		7½
Number of cells	120		250		660		4,510		14,290
Number of divisions		1.04		1.32		2.71		1.58	
Mean cell cycle (hr)		11.5		9.1		4.4		6.7	

Source: Snow 1977.

1977). Six and a half days after fertilization is also the time that the primitive streak appears, marking the onset of axis formation and bilateral symmetry in the embryo.

A study of cell numbers and mitotic activity in embryos at the primitive streak stage led Snow (1977) to the realization that the rapid mean rate of cell division was not distributed evenly over the egg cylinder: a small region, which he termed the proliferative zone, opposite the primitive streak and constituting about 10 percent of the epiblast, appears to have a cell generation time of less than 3 hours over a 24-hour period and in this time generates about half the cells in the 7½-day embryo. Evidence from an analysis of t^{w18} mutant embryos (Snow and Bennett 1978) suggests that the cells generated in the proliferative zone give rise mainly to ectoderm, while the rest of the epiblast cells pass through the primitive streak and become mesoderm.

No primordial germ cells can yet be seen. The earliest germ cells to have been detected in a mouse embryo were located at the base of the allantois 8 days *post coitum* (Ożdżeński 1967); they were identified as primordial germ cells by their high alkaline phosphatase activity. At this stage, the embryo has a head fold and a tail bud, and the first of the somites is about to form.

Whose genes in control?

In Amphibia, the embryo's own genes only begin to control development at around the time of gastrulation. If this were true of the mouse also, almost all the development that has been described so far would have been directed by the maternal genome, through messenger RNA stored in the egg cytoplasm. What is the evidence for the mouse embryo? Whose genes are in control?

The unfertilized mouse egg contains a large amount of messenger RNA, much of which may be stored in some masked form that is not yet involved in transcription. Between the unfertilized egg and the embryo at the early cleavage stage, qualitative changes are seen in the pattern of polypeptides synthesized. Some of these, at least in the rabbit, are merely temporal changes—chronological changes—occurring at the same time whether or not the egg is fertilized (Van Blerkom 1979). Since transcription is unlikely to occur in an unfertilized egg with the chromosomes blocked in second meiotic metaphase, these changes are presumably due to post-transcriptional modification. Other changes in the pattern of polypeptide synthesis are sparked off only by fertilization or activation. For some of these also, Braude and his colleagues (1979) have evidence that the increased synthesis is not dependent on transcription but represents post-transcriptional control of messenger RNA synthesized before fertilization. If the messenger population of the unfertilized egg is extracted and translated in a cell-free *in vitro* system, the same polypeptides appear as would normally first appear at the 2-cell stage. In the intact unfertilized egg, the relevant messenger RNA molecules must therefore be present in an inactive, masked form, only to be used as the embryo begins to develop.

Two recent studies suggest that some maternal message persists for the first few cleavage divisions, but most is degraded by the blastocyst stage. In the first study (Bachvarova & De Leon 1980), the RNA in the unfertilized egg was labelled by exposing oocytes to radioactive uridine during the period of oocyte growth. Of the labelled maternal RNA, 40 percent was degraded on the first day after fertilization, before the 2-cell stage; none was degraded on the second day; and a further 30 percent was degraded on the third day, at the 8-cell stage. A similar time scale emerges from an experiment in which mouse or rabbit globin message was injected into fertilized mouse eggs (Brinster et al. 1980). Globin synthesis could be detected up to 24 hours after injection, but not at the blastocyst stage, and probably not at the 8-cell stage. This result is in striking contrast to the situation in the *Xenopus* oocyte, where an injected globin message retains its activity for at least 2 weeks.

So if maternal message transmitted in the egg cytoplasm persists and perhaps directs development up to the 8-cell stage but rarely beyond the blastocyst stage, when is the embryonic genome first expressed? Recent reviews (for example, McLaren 1976a; Sherman 1979) summarize much evidence suggesting that the embryonic genome is not only transcribed but also expressed during early cleavage, possibly at the 2-cell stage, certainly by the 8-cell stage. Messenger, ribosomal, 4S, and 5S RNA are all synthesized at the 2–4 cell stage. Paternally inherited variants of several enzymes and antigens can be detected at the 8-cell stage (table 2). At least three genetic factors that are lethal in homozygous form, namely, the genes *agouti yellow* and t^{12} and the albino locus deletion c^{25H}, are first expressed at the 2-cell stage. Inhibitor studies tell the same story. Exposure to α-amanitin can block development as early as the

Table 2. Embryonic Gene Activity by the 8-Cell Stage

RNA
- Ribosomal
- Heterogeneous
- 4S, 5S

Cell surface antigens

Enzyme variants
- β-glucuronidase
- GPI
- PGD (rat)

Recessive lethal genes
- A^y
- t^{12}

Chromosome effects
- c^{25H} (deletion)
- X-chromosome activity

2-cell stage. Blastocyst formation can be prevented by exposure to α-amanitin up to about 68 hours after fertilization but not thereafter, suggesting that a critical transcriptional event takes place at this time (Braude 1979).

X-chromosome function

A particularly revealing picture emerges from recent studies on X-chromosome function. In our own species, the X-chromosome-coded dimeric enzyme glucose-6-phosphate dehydrogenase (G6PD) has been used to show that both X chromosomes are transcriptionally active during the development of the oocyte. Oocytes from women who are heterozygous for a genetically determined electrophoretic variant of G6PD show not only the two bands corresponding to the two pure isozymes but also the hybrid band expected from a dimer if both subunits are synthesized in the same cell (Gartler, Liskay, and Gant

1973). This is good evidence that both X chromosomes are switched on in oocytes. In mice we have as yet no X-chromosome-coded dimer, so we have to look at activity levels of enzymes. If we compare oocytes from XO mice, possessing only a single X chromosome, with oocytes from normal XX mice, those from XO mice show only half as much activity for enzymes coded by the X chromosome as do those from XX females; see Epstein (1969, G6PD; 1972, hypoxanthine phosphoribosyl transferase, HPRT) and Kozak and Quinn (1975, phosphoglycerate kinase, PGK). Autosomally coded enzymes of course show no such effect. The two-fold difference in enzyme activity persists through fertilization and early cleavage up to the 8-cell stage, at least for the enzyme HPRT (Monk and Harper 1978). After compaction has occurred, the twofold difference disappears. This suggests that enzyme activity up to the 8-cell stage is controlled by maternal gene products—probably messenger RNA stored in the egg cytoplasm rather than preformed protein precursors, since HPRT activity during this period increases some fifteenfold.

Not surprisingly, there are differences in the development of embryos from XX and XO mothers which reflect the difference in enzyme activity. Embryos from XO mothers, irrespective of their own chromosome constitution, show an abnormally high incidence of developmental arrest in mid-cleavage (Burgoyne and Biggers 1976), presumably because they are deficient in X-coded molecules synthesized during oocyte growth.

If we look at embryos from normal XX females at the post-compaction, morula stage, we see for the first time a bimodal distribution of enzyme activity in individual embryos, not only for HPRT (Monk and Harper 1978; Kratzer and Gartler 1978) but also for another X-coded enzyme, α-galactosidase (Adler, West, and Chapman 1977). The two modes differ in activity by a factor of up to two, and chromosome determinations have shown the higher activity levels to be present in XX, the lower in XY embryos (Epstein, Smith, Travis, and Tucker 1978). These results establish that both X chromosomes are switched on in female embryos at this time and confirm that the embryonic genome is expressed by mid-cleavage.

XO embryos, even those derived from XX mothers and hence not handicapped by the maternal effect suffered initially by any embryo developing from an XO oocyte, might be expected to show some abnormalities during the period when both X chromosomes are active in XX embryos. Burgoyne (personal communication) has indeed observed that XO embryos from normal XX mothers tend to arrest at the morula/blastocyst stage.

Late blastocysts no longer show a bimodal distribution of HPRT activity, which suggests that one X chromosome in XX embryos has already been inactivated. On the other hand, Gardner and Lyon (1971) have convincing genetic evidence that X-inactivation has not occurred by this stage in the ancestors of coat color pigment cells. The apparent paradox has been resolved by the demonstration that a bimodal distribution of HPRT activity can still be found in the inner cell mass of the blastocyst, once the numerically predominant trophectoderm cells have been removed, and in the epiblast derived from the inner cell mass at 6 days *post coitum* (Monk and Harper 1979). Chromosome studies of the 6-day embryos have again confirmed that the higher values are derived from XX and the lower from XY embryos. Other 6-day tissues that have already differentiated (proximal endoderm, extra-embryonic ectoderm) show a unimodal distribution of HPRT activity, as does the epiblast at 6½ days *post coitum*. Thus by the time that primitive-streak formation begins X-chromosome inactivation seems to have occurred throughout the conceptus.

X-chromosome inactivation has also been examined in an XX teratocarcinoma (embryonic tumor) cell line maintained *in*

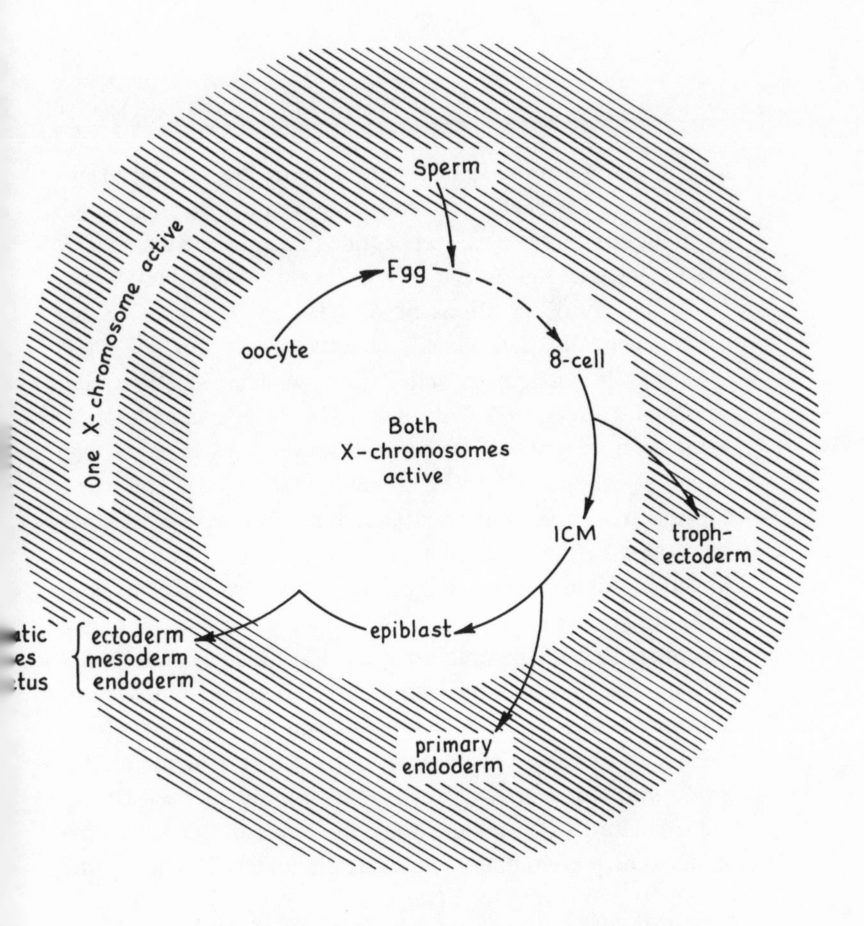

Figure 13. X-Chromosome Activity from Oocyte to Gastrula. In XX embryos, one X chromosome appears to become inactivated in tissues undergoing primary differentiation, while both X chromosomes remain active in the stem-cell line linking oocyte to epiblast. (After Monk and Harper 1979, figure 1).

vitro (Martin et al 1978). As in the embryo, the undifferentiated stem cells have two X chromosomes active, the differentiated tissues only one. The link between X-chromosome inactivation and the restriction of developmental potency (see figure 13) has been stressed by Monk (1978).

In the embryo, the derivatives of tissues that differentiate early and undergo X-inactivation early—that is, the trophectoderm and the primary endoderm—show preferential inactivation of the paternally derived X chromosome, as judged both by studies of chromosome replication (Takagi and Sasaki 1975; Wake, Takagi, and Sasaki 1976; Takagi, Wake, and Sasaki 1978) and by observations of X-coded enzyme activity that used genetically determined variants of the enzyme PGK (West, Frels, Chapman, and Papaioannou, 1977; West, Papaioannou, Frels, and Chapman 1978). The nonrandomness seems to be a function of the inactivation process rather than of cell selection (Takagi, Wake, and Sasaki 1978). The paternal X is not intrinsically nonfunctional in these tissues, since expression of the paternal variant of PGK has been demonstrated in trophectoderm and primary endoderm derivatives of XO embryos whose single X is of paternal origin (Frels and Chapman 1979). The epiblast is the last tissue to undergo X-chromosome inactivation, and maternal and paternal X chromosomes now seem equally at risk (Lyon 1972), as though the passage of time has erased some trace of their gametic origin. (For a discussion of X-chromosome "imprinting," see Brown and Chandra 1973). In kangaroos, the paternal X chromosome seems to be preferentially inactivated in all tissues, embryonic as well as extraembryonic (see Cooper et al. 1975). In diploid parthenogenetic mouse embryos, where both X chromosomes are derived from the mother, one X is still inactivated (Kaufman, Guc-Cubrilo, and Lyon 1978), but we do not yet know whether the same is true of extraembryonic tissues where the inactive X would normally be derived from the father.

5 Commitment

So far we have dealt only with the normal course of development—what happens and when—at the morphological, cellular and, to a small extent, the biochemical, level. But suppose we perturb development. What powers of regulation does the early embryo possess, and what information can we derive about the developmental potential of the various tissues and the time at which they become committed?

Time of commitment

Up to the 8-cell stage, perhaps up to the time of compaction, the powers of regulation of the mammalian embryo seem total. Two- or 8-cell mouse embryos aggregated together will form a single chimeric individual; single blastomeres isolated from 8-cell rabbit or sheep embryos or 2-cell mouse embryos develop into normal adults, and single 8-cell mouse blastomeres fail at the blastocyst stage only because the total number of cells at that stage is too small to permit embryogenesis. Throughout this period each blastomere seems totipotent: an 8-cell cow embryo split into 4 pairs of identical quadruplets (Willadsen and Polge 1981), and all four 4-cell- and most 8-cell-stage mouse blastomeres, when aggregated with carrier blastomeres of distinguishable genetic type, have been shown to give rise to both trophectoderm and inner cell mass derivatives (Kelly 1977).

On the other hand, we have seen in chapter 4 that in normal development allocation takes place at a very early stage, so that, for example, the first blastomere to divide from the 2-cell stage contributes disproportionately to the inner cell mass. If this early allocation were due to cytoplasmic localization of trophectoderm and inner-cell-mass determinants in the fertilized egg, it would be hard to reconcile with the powers of regulation that exist at the same period. In fact, however, it reflects the regularity of the normal cleavage pattern. It seems that the fate of a blastomere (trophectoderm versus inner cell mass, also probably endoderm versus epiblast at a later stage) depends not on its origin but rather on its position within the embryo, so that interior cells at the 12–16 cell stage become inner cell mass and peripheral cells become trophectoderm. If at the 8-cell stage the pattern is disturbed by reaggregation of blastomeres, a cell placed on the periphery of the cell mass is likely to contribute descendants only to the trophectoderm, while one wholly surrounded by other cells will tend to contribute at least some descendants to the inner cell mass (Hillman, Sherman and Graham 1972). If too few cells are present at the critical stage, all take the trophectoderm pathway, and no inner cell mass develops.

Up to the early blastocyst stage, the inside cells retain developmental lability and can regenerate trophectoderm if they are isolated (Handyside 1978; Spindle 1978). The inner cell mass from a mature blastocyst, on the other hand, never gives rise to trophectoderm or trophectoderm derivatives, no matter whether it is isolated in culture (Handyside and Barton 1977) or aggregated with an 8-cell embryo (Rossant 1975). Similarly, outside cells recovered from late morulae and aggregated with 8-cell embryos are capable of contributing to both inner cell mass and trophectoderm, and morulae reconstituted entirely from cells that were previously outside give rise to morphologically normal embryos (Rossant and Vijh 1980). In the mature blastocyst, however, the trophectoderm appears to be irreversibly committed, and trophectoderm cells have never been induced to contribute to inner cell mass derivatives (Gardner and Johnson 1972; Gardner and Rossant 1976).

At the biochemical level too, irreversible commitment does not take place until the late blastocyst stage. Whereas the inside cells of the late morula synthesize not only inner-cell-mass-specific but also at least one trophectoderm-specific polypeptide, in the mature inner cell mass no trophectoderm markers can be detected (Handyside and Johnson 1978). If an inside cell from a late morula that is already allocated to the inner cell mass and is synthesizing inner-cell-mass-specific polypeptides is isolated so that it switches and becomes committed to a trophectoderm pathway, the pattern of polypeptide synthesis shifts, and trophectoderm-specific polypeptides begin to be synthesized (Johnson 1979). It seems that differentiation takes place earlier than commitment. In the presence of α-amanatin, the isolated inner cell masses no longer form trophectoderm, and no shift in polypeptide synthesis pattern can be detected; so the positional response must involve transcriptional control—in other words, some shift in gene expression. Inner cell masses from mature blastocysts, which do not give rise to trophectoderm when isolated, nonetheless show some changes in the pattern of polypeptide synthesis in a trophectodermlike direction, as though they were giving a subthreshold, morphologically ineffective response to their altered position.

The timing of commitment may be subject to genetic variation, since Hogan and Tilly (1978) claim that inner cell masses from mature blastocysts of the C3H strain of mice will give rise to giant trophoblast cells if they are allowed to develop into egg-cylinderlike structures *in vitro* and are then disrupted and cultured further. This suggests that commitment is delayed on a C3H genetic background. An alternative explanation, put forward by the authors, is that some of the extraembryonic ectoderm in the normal embryos is derived from the inner cell mass, rather than all from the trophectoderm, as previously proposed (Gardner, Papaioannou, and Barton 1973), and that the giant trophoblast cells in their cultures are derived from this extraembryonic ectoderm. Another possibility is that the giant cells observed were not trophoblast cells at all but belonged to some other lineage, for example, endoderm.

"Inside-outside" differentiation

The "inside-outside theory" (Graham 1971) provides a causal basis for trophectoderm/inner-cell-mass differentiation but leaves open the precise cellular mechanisms that distinguish the interior from the periphery. It has been suggested that the permeability seal provided by the peripheral tight junctions makes differentiation possible by creating a chemically distinct interior microenvironment (Ducibella et al. 1975). In fact, this now seems not to be so. When compaction, and hence junction formation, are prevented by culturing embryos in the presence of either antiserum (Johnson et al. 1979) or cytochalasin D (H. Pratt, personal communication), polypeptides characteristic of the inner cell mass are still synthesized on schedule. So an enclosed interior microenvironment seems not to be a necessary prerequisite, at least for the initial biochemical changes associated with differentiation. These changes may be genetically programmed and, in the absence of positional inside/outside information, may occur in all the cells of the embryo. The timing of these changes is still obscure but, as with blastocoel formation (Smith and McLaren 1977), does not depend on cell number or cytokinesis. The same series of experiments suggests that compaction is required for intercellular heterogeneity to be generated within the embryo. Johnson and his colleagues suggest that the positional information required for inside/outside differentiation is derived from the intercellular interactions and cell polarization that first become obvious at the 8-cell stage. The development of surface polarity at this time seems to be closely associated with cell contact (Ziomek and Johnson 1981).

If the block to compaction is removed during the period when the inner cell mass is normally still labile, compaction and blastocyst formation take place within a few hours, and subsequent development is not affected. If the block is extended over a longer time into the period when the blastocyst would normally be expanding, however, its removal is followed by abnormal development. This is also the period of development, as Johnson (1979) showed, when isolated inner cell masses, though by now incapable of differentiating as trophectoderm, nonetheless undergo some changes of polypeptide synthesis pattern in a trophectodermal direction. These two experiments—the first taking normal inner cell masses and exposing them to an outside environment, the second delaying the formation of a normal inside environment until after the normal time of blastocyst formation—are complementary, and both suggest that determination involves a gradual shift in biochemical profile, rather than an all-or-nothing switch mechanism. As Johnson puts it, his results support "models for commitment which do not invoke a special, quantal molecular mechanism but rather a progressive accumulation of differentiative changes."

Developmental potency

At about the same time that its capacity to generate trophectoderm is lost, the inner cell mass begins to differentiate primary endoderm. In a normal embryo, the cells that take the endodermal pathway are those adjacent to the blastocoel cavity, which comprise about 50 percent of the total inner cell mass. In an inner cell mass removed from its trophectoderm shell, all the peripheral cells differentiate as endoderm. If the total number of cells is low, this may include every cell in the inner cell mass.

Up to the early egg cylinder stage, the developmental potency of various embryonic and extraembryonic tissues can be assessed by injecting isolated cells into a 3½-day host blastocyst of contrasting genetic type and determining by subsequent analysis to which tissues their progeny contribute. When primary endoderm cells are removed from a blastocyst 4½ days after fertilization and are injected into a host blastocyst, their progeny are later found in the endoderm layer of the visceral yolk sac, not in the yolk-sac mesoderm nor in the fetus, not even in the endoderm of the fetus (Gardner and Rossant 1979). This result suggests that endoderm is committed to its developmental pathway more or less as soon as it is first distinguishable. It also confirms the widely held view that the definitive fetal endoderm arises from the primitive streak and is not derived from primary endoderm. Epiblast cells from a blastocyst of the same age will colonize the fetus and the yolk-sac mesoderm but will never contribute to the yolk-sac endoderm. Epiblast in culture, in the absence of endoderm, may retain its power to regenerate an endodermal layer longer, much as inner cell mass retains for some time its power to regenerate trophectoderm if it is isolated.

Twenty-four hours later—5½ days after fertilization—proximal endoderm cells taken from either embryonic or extraembryonic regions of the egg cylinder and injected into a host blastocyst also contribute progeny to the yolk-sac endoderm, but not to the yolk-sac mesoderm or the placenta, nor to any fetal tissues, endodermal or otherwise (Rossant, Gardner, and Alexandre 1978). Extraembryonic ectoderm cells taken either 5½ or 6½ days after fertilization and injected into a host blastocyst colonize the ectoplacental cone and trophoblast giant cell populations but do not contribute to the embryo or the yolk sac.

At later stages, the developmental potency of embryonic tissues can no longer be assessed by transfer to a blastocyst, but some information is available from experiments where rat embryonic tissues, alone or in combination, were transferred beneath the kidney capsule (Skreb, Svajger, and Levak-Svajger 1976). Endoderm from a two-layered egg cylinder fails to grow, while ectoderm gives rise to ectodermal, mesodermal, and endodermal derivatives. From a three-layered egg cylinder, taken one day later, ectoderm still gives rise to ectodermal and mesodermal derivatives, but no longer to endodermal derivatives; endoderm on its own fails to grow, and mesoderm produces only brown adipose tissue. Endoderm and mesoderm together, on the other hand, give rise to a wide range of endodermal and mesodermal derivatives, suggesting an inductive relationship.

The mouse egg cylinder, at around the time of gastrulation, shows remarkable powers of regulation (table 3). Treatment with the DNA-synthesis inhibitor mitomycin C *in vivo* or *in vitro* can reduce cell number to as little as 15 percent of the control level without killing the embryo, and compensatory growth makes up the loss in cell number within a few days

Table 3. Cell Numbers and Mitotic Activity in Control and MMC-Treated 7½-Day Embryos (MMC injected 7.0 days *p.c.*)

	Tissue	Control	MMC
No. of cells	Ectoderm	7,618 ± 443	978 ± 157
	Mesoderm	3,743 ± 256	374 ± 53
	Endoderm	1,019 ± 57	374 ± 33
Mitotic index	Ectoderm	3.9 ± 0.2	12.5 ± 1.5
	Mesoderm	1.9 ± 0.2	5.2 ± 0.7
	Endoderm	4.5 ± 0.6	6.1 ± 0.7

Note: MMC = Mitomycin-C. Camera lucida drawings were made of transverse sections and tissue volumes were computed from planimeter measurements of tissue areas. Cell numbers were then calculated from a knowledge of cell volume.
Source: Snow and Tam 1979.

(Snow and Tam 1979). Yet, paradoxically, this regulative power at the level of the whole embryo is not reflected in any regenerative capacity at the level of its component parts. If small fragments of tissue are removed from a mouse egg cylinder and are cultured for 24–48 hours, they give rise to the same structures that they would have produced *in situ*. The depleted donor egg cylinder continues to develop *in vitro* but does not replace the missing parts (Snow, personal communication). Thus, by the time of gastrulation, the developmental potency of most parts of the embryo would appear to have become markedly restricted, in the sense that when isolated they do not regenerate adjacent structures. This means that it is now possible for the first time to construct a preliminary fate map for the normal primitive-streak stage mouse embryo (figure 14). Whether the developmental fate of the various parts could be altered by grafting them into different regions of the embryo remains to be explored.

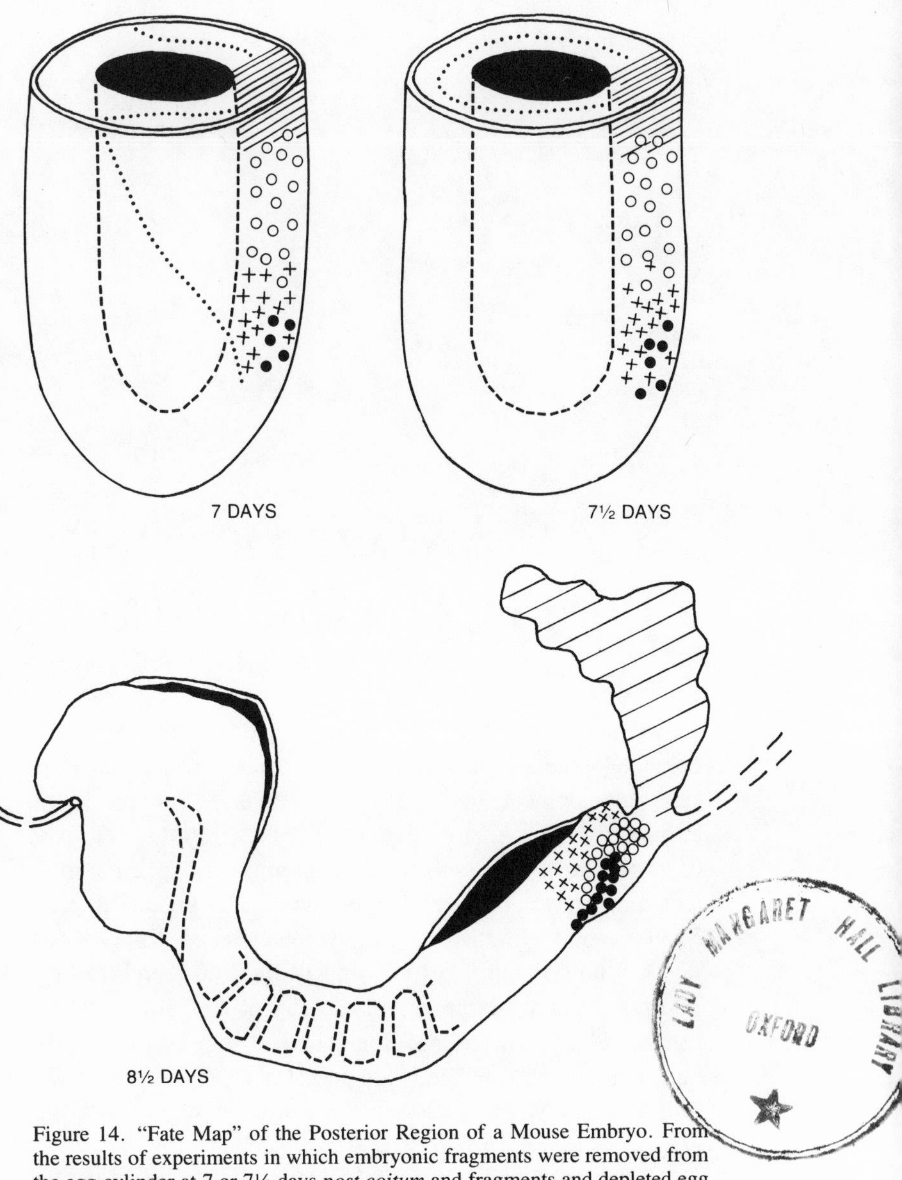

Figure 14. "Fate Map" of the Posterior Region of a Mouse Embryo. From the results of experiments in which embryonic fragments were removed from the egg cylinder at 7 or 7½ days *post coitum* and fragments and depleted egg cylinders were cultured for 24–36 hours. The primordial germ cells were identified by alkaline phosphatase content. ○ hind gut endoderm ● primordial germ cells × tail bud //// allantois (After Snow, personal communication.)

6

Where Do All the Germ Cells Come From?

As we have seen, a certain amount is known about the developmental potency of various cell types at various stages of early development; but mostly this knowledge is expressed in terms of whether a particular cell gives rise to inner cell mass or trophectoderm or both, or to embryonic or extraembryonic structures, or to ectodermal or mesodermal or endodermal derivatives in the fetus. Surgically manipulated embryos are rarely followed through to postnatal life, and therefore little information is available on the ancestry of the germ cells. We cannot yet say, for any developmental stage prior to gastrulation, that one or a few cells are ancestral to all the germ cells. Still less can we say that one or a few cells give rise only to germ cells and not to any somatic cells. For all we know, all cells in the embryo may contribute both to the the germ-cell and to the somatic-cell population.

We know that in the mouse all four blastomeres at the 4-cell

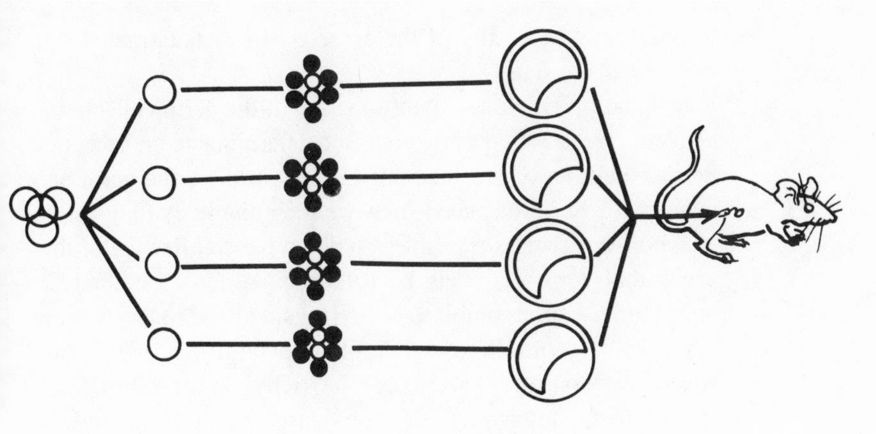

Figure 15. Design for the "Quartet" Experiments of Kelly (1977). The donor embryo (white) is obtained at the 4-cell stage, the zona pellucida is removed, and the blastomeres are dissociated and cultured. At the 8-cell stage, the blastomeres are aggregated with carrier blastomeres (black); and at the blastocyst stage, the aggregates are transferred to pseudopregnant recipients. Three of the set of four mice from two quartets survived and bred, and all three had germ cells of the donor (white) genetic type. (From Kelly 1977, figure 1.)

stage are totipotent in the sense that they give rise to both inner cell mass and trophectoderm, and we also know (Kelly 1977) that at least three out of these four are capable of giving rise to germ cells (figure 15). This implies that both blastomeres at the 2-cell stage can give rise to germ cells, a result that had been strongly suggested by an earlier finding (Tarkowski 1959), that six mice developed from 2-cell stages of which one blastomere had been destroyed were all fertile. Now that several pairs of monozygotic twin lambs (Willadsen 1978) and one set of quadruplet calves (Willadsen and Polge 1981) have been made by splitting embryos at the 2- and 8-cell stages, we should soon have similar information for sheep and cow embryos. It will be surprising if the germ line turns out to reside in one blastomere only of the embryo of any mammal at the early cleavage stage.

In some invertebrates, the formation of the germ cells is dependent on the action of cytoplasmic determinants present earlier, at the time of fertilization. In such species, the germinal cytoplasm, or germ plasm, may be identifiable cytologically, which means that its transmission down the cell lineage of the primordial germ cells can be followed. Among vertebrates, only the anuran amphibians—the frogs and the toads—show evidence of germ plasm (see Smith and Williams 1979). The areas of cytoplasm involved are very rich in mitochondria and also contain numerous electron-dense germinal granules, thought to be made up of ribonucleoprotein. No evidence of germ plasm has been found in urodeles or birds. In mammals, the germ cells themselves, as we shall see in the next chapter, are often characterized by electron-dense bodies known as nuage, and it has been suggested that this nuage may be equivalent to the anuran germ plasm. However, there are no reports of such material in the early mammalian embryo before the emergence of germ cells, and the only localized concentrations of mitochondria to be seen are those adjacent to the apposed cell membranes at the compacted 8-cell stage.

At the blastocyst stage, injections into host blastocysts of single cells taken from the inner cell mass 3½ days after fertilization or from the epiblast 4½ days after fertilization have yielded mice that showed both somatic and germ-line chimerism (Gardner 1977 and personal communication). Of those coat color chimeras in which both donor and host components were of the same sex, so that both might have been expected to give rise to germ cells, 4 of 6 from the 3½-day and 2 of 7 from the 4½-day group proved in breeding tests to be germ-line chimeras (table 4). This result establishes that germ cells derive from the epiblast and not from the primary endoderm.

Table 4. The Incidence of Germ-Line Chimerism in Coat Color Chimeras with Both Components of the Same Sex

	Germ-line chimerism		
Donor cell	Present	Absent	Total
Days *post coitum*			
3½	4	2	6
4½	2	5	7
Total	6	7	13

Note: Single cells from 3½-day inner cell masses or 4½-day epiblasts were injected into 3½-day mouse blastocysts.

Source: Data from Gardner 1977 and personal communication.

We have already seen that epiblast cells injected into blastocysts do not contribute to endodermal derivatives; nor do they give rise to trophectoderm derivatives, even when they are isolated in culture; so the fact that they give rise to germ cells means that germ cells, though themselves totipotent, are derived from cells of restricted potency. Gardner's result also establishes that individual donor cells can contribute to both somatic- and germ-cell populations. Since the 3½-day inner cell mass is likely to contain about 15 cells and the 4½-day epiblast at least 20 cells (McLaren 1976b), the relatively high proportion of germ-line chimeras suggests that germ cells are derived from more than one—perhaps from the majority—of the cells present at each stage. However, the very uneven proliferation rates shown by different cell populations in the egg cylinder (Snow 1977) make such statistical inferences somewhat questionable.

The first stage at which we can be reasonably confident that more than one cell in the normal, unperturbed embryo will give rise to germ cells is the 5½-day egg cylinder in which the proamniotic cavity is just forming. This is the stage at which aggregation chimeras shift from being twice the size of control embryos to being the same size (Buehr and McLaren 1974). We know that both components of a chimera can contribute to the germ-cell population, so in the chimera there must be at least two germ-cell progenitors present at every stage. Once size regulation has taken place, provided that all cell populations have been reduced in proportion, the chimera will contain the same number of germ-cell progenitors as would a control embryo. It follows, therefore, that control embryos at this stage also contain at least two germ-cell progenitors.

Within the next 24 hours, the epiblast, as we have seen, appears to undergo X-chromosome inactivation. Does this apply to the germ-cell progenitors, as well as to the rest of the epiblast, or is a small germ-cell progenitor population with two active X chromosomes concealed within a larger one-active-X population, in the same way that the two-active-X-chromosome status of the inner cell mass is masked in the intact blastocyst by the numerically greater trophectoderm population? Does the female germ-cell lineage, as part of its totipotency, keep both X chromosomes switched on throughout its life cycle? We shall look further at this question in chapter 8.

We have now again reached the stage of gastrulation, without finding any trace of a distinctive germ-cell lineage. In chapter 5 we saw that the recent work of Snow makes possible the construction of a preliminary fate map of the mouse embryo at the primitive-streak stage by removing fragments of tissue from the egg cylinder and culturing them to see what structures they yield. One fragment that has been removed is a small piece at the hind end of the primitive streak of a 7-day embryo. After 24 hours of culture (figure 16), this fragment consistently forms primordial germ cells—large, round, alkaline-phosphatase-positive germ cells just like those found by Ożdżeński (1967) in the intact 8-day embryo at the base of the allantois. Not only has this fragment formed germ cells, but it has formed almost as many germ cells as would have formed

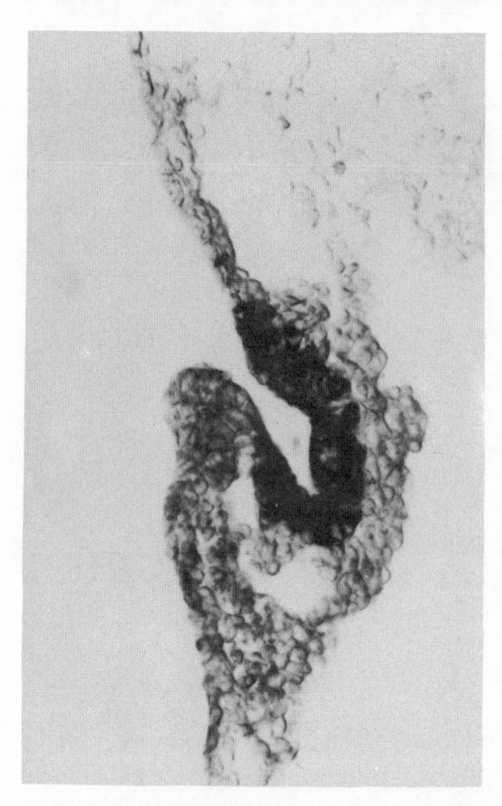

Figure 16. Embryonic Fragment Excised from the Posterior Part of the Egg Cylinder at 7½ Days *post Coitum*, Cultured for 24 Hours, and Stained for Alkaline Phosphatase Activity. The primordial germ cells, showing high alkaline phosphatase activity, are clustered around the hind gut. (From Snow, personal communication.)

in the intact embryo, and the depleted donor egg cylinder turns out to be almost entirely devoid of germ cells (Snow, personal communication).

So in a small fragment of tissue containing only one or two hundred cells, we see the sudden emergence of the next generation of germ cells. We know that their ancestors were present in the same fragment of tissue 24 hours earlier—that is, 7 days after fertilization—but we do not yet know whether they were brought there by the backward movement of the primitive streak or whether they, and their ancestors before them, have resided in that corner of the epiblast, near the extraembryonic ectoderm, since a much earlier stage of development.

What happens to these germ cells next? In the next two chapters, we shall follow them in their migration to the genital ridges and see first how they affect sex determination, and then how they in their turn respond to sex determination.

7 Migration

Emergence

The first person to identify primordial germ cells in the early mouse embryo positively, using selective staining for their high alkaline phosphatase activity as a marker, was Chiquoine in 1954. He described the appearance of germ cells in 8-day embryos at three sites: the caudal end of the primitive streak, the mesoderm at the root of the allantois, and the underlying endoderm of the yolk sac. He judged that they had all originated in the yolk sac, since that was where they were most abundant in those embryos that contained fewest germ cells. This conclusion was upheld by Mintz and Russell (1957). Ożdżeński (1967), on the other hand, examined embryos at a slightly earlier stage—before any somites or head fold had formed—and found that virtually all the primordial germ cells were in the embryonic part of the allantoic rudiment, which represents an extension of the primitive streak. Only later were they seen at the end of the primitive streak itself and in the yolk-sac endoderm. In partial confirmation, the fine structural observations of Clark and Eddy (1975) showed that the primordial germ cells of the 8-day embryo were morphologically more similar to mesoderm or ectoderm cells than to endoderm.

Can we be sure that the primordial germ cells identified in these studies are indeed the ancestors of the definitive germ cells of the adult gonad? The first convincing demonstration was provided by Mintz and Russell (1957). In embryos homozygous for *W* genes, where adults are sterile because the gonads are devoid of germ cells, the primary defect turned out to be failure of the primordial germ cells to proliferate normally.

Snow (personal communication) recently discovered that if a small piece of tissue excised from an area near the caudal end of the primitive streak on the day before any primordial germ cells are apparent is cultured in isolation for 24 hours, it develops almost the entire complement of germ cells that would have been found in an intact embryo. At the time of its explantation, this fragment of tissue contains only about 150–200 cells; what proportion of these are progenitors of germ cells we do not yet know. *In situ*, this region moves in an extraembryonic direction during the next 24 hours, owing perhaps to the pressure of cell division in the neighboring regions. The allantoic rudiment is extruded, and the primordial germ cells first reveal themselves, by their high alkaline phosphatase activity, at its base.

Route of migration

The subsequent migration route of the primordial germ cells of the mouse (Chiquoine 1954; Mintz and Russell 1957; Spiegelman and Bennett 1973) proves to be very similar to that described by Witschi (1948) for the human embryo. From their initial caudal position, the primordial germ cells move into the wall of the hind gut, both in the endoderm and in the mesoderm. This may not involve any active movement, since at this time the hind gut is invaginating and pulling into itself much of this caudal area. Clark and Eddy (1975) point out that at this stage the germ cells do not yet possess the ultrastructural features of locomotory cells. Once in the hind gut, however, they change in appearance and begin an active migration through the solid tissue of the gut wall. Within a couple of days they leave the gut through the dorsal mesentery and travel around the dorsal aorta and the coelomic angles into the genital ridges, the site of the future gonads (figure 17). Some germ cells wander off and are seen at such sites as the skin ectoderm, the allantois, the mesonephric tubules, and the mesenchyme near the neural tube; these misplaced germ cells presumably never succeed in reaching the genital ridges. When Snow's cultured fragments include the area that invaginates to form the hind gut, the germ cells congregate within its walls (see figure 16); but when no hind gut is present, they disperse widely (figure 18).

Much debate has centered around the question of whether primordial germ cells in the mammal are ever transported passively in the bloodstream from their place of origin to the genital ridges, as they are in the chick. Cells thought to be primordial germ cells have been reported within embryonic blood vessels in the calf (Ohno and Gropp 1965), the human, and the rat (Semyonova-Tian-Shanskaya 1976), but there is no means of telling whether such germ cells—if that is indeed what they were—would have ended up in the gonads. Breeding studies on cattle and marmosets, where twin embryos have a common blood circulation, have failed to yield any convincing evidence for exchange of germ cells (see Ford and Evans 1977), though germ-cell mosaicism in the gonads of cattle twins has been

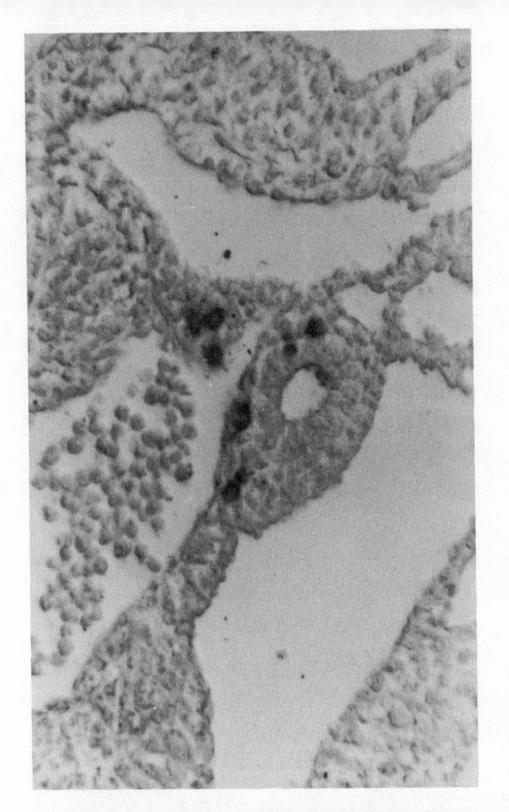

Figure 17. Transverse Section of Mouse Embryo 9½ Days *post Coitum*, Stained for Alkaline Phosphatase Activity. The primordial germ cells, which show high alkaline phosphatase activity, are migrating from the hind gut up the dorsal mesentery and into the genital ridge. (From Tam and Snow 1980, figure 3d.)

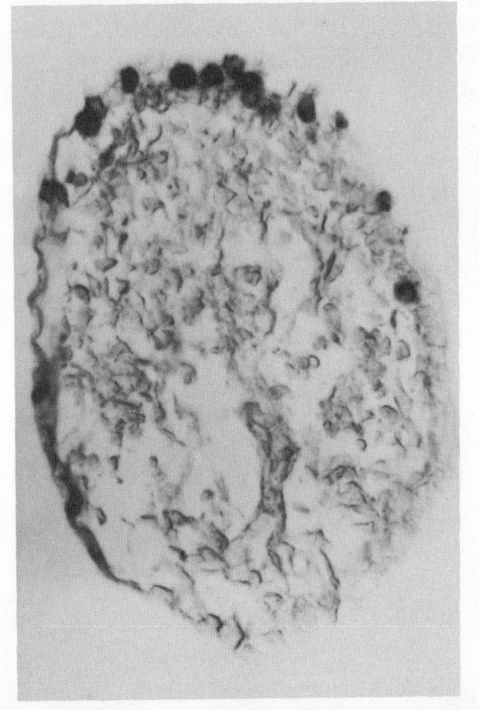

Figure 18. Embryonic Fragment Excised from the Posterior Part of the Egg Cylinder at 7½ Days *post Coitum*, Cultured for 24 Hours, and Stained for Alkaline Phosphatase Activity. This fragment contains no hind gut tissue (see figure 16), and primordial germ cells, which show high alkaline phosphatase activity, have migrated to the periphery of the allantoic bud. (From Snow, personal communication.)

claimed on cytological grounds (Ohno et al. 1962). The case for a vascular migration route in the mammal thus remains unproven.

How then do the primordial germ cells move through the solid tissues of the gut wall and dorsal mesentery? Heasman, Mohun, and Wylie (1977) analyzed by time-lapse cinematography the movement of primordial germ cells of *Xenopus* over various substrates, and concluded that locomotion *in vivo* probably takes place by elongation coupled with the extrusion of filopodia, followed by waves of contraction rather than by amoeboid movement. The migration of chick primordial germ cells can be inhibited by treatment of the early embryo with concanavalin A: the affected germ cells become more rounded and no longer show the cell surface "ruffling" associated with locomotor activity (Lee, Karasanyi, and Nagele 1978). Pseudopodia in mouse primordial germ cells have been detected histologically by Chiquoine (1954) and Clark and Eddy (1975), while Blandau, White, and Rumery (1963) have described and filmed the amoeboid locomotory movements of mouse germ cells isolated from the hind gut and genital ridges. Female germ cells showed locomotory behavior up to the pachytene stage of meiosis, but male germ cells became quiescent as soon as the testis cords had formed. In view of the findings of Heasman, Mohun, and Wylie (1977), it is possible that the type of locomotion described by Blandau, White, and Rumery may have been affected by the noncellular substrate used (glass) and may not fully reflect locomotory behavior *in vivo*.

Route-finding—how the germ cells find their way—remains deeply mysterious. For the journey along the gut, the germ cells are all confined to the epithelium by a basal lamina along the outer surface and extensive junctions at the lateral margins of the cells, just below the inner, luminal surface (Clark and

Eddy 1975). Primordial germ cells of *Xenopus*, which also migrate through solid tissues *in vivo*, invade a cellular substrate *in vitro*, sending out broad processes that penetrate between the underlying cells (Heasman and Wylie 1978). Wylie and co-workers (1979) found that *Xenopus* primordial germ cells use a contact guidance mechanism *in vitro* and suggest that migrating germ cells *in vivo* follow ridges and striations on the underlying coelomic lining cells. Some chemical that attracts the cells, like cyclic AMP in slime molds, seems to provide a likely explanation for the movement of the primordial germ cells out of the wall of the hind gut and into the genital ridges and perhaps also for their initial aggregation into the hind gut, unless this is brought about by morphogenetic movements. A chemotactic attraction exerted by the genital ridges on the migrating primordial germ cells has been shown to occur in the chick (see Dubois 1968; Dubois and Croisille 1970). When pieces of mouse hind gut containing germ cells were transplanted into the coelomic cavity of chick embryos, some of the mouse germ cells settled in or near the chick gonads and mesonephros, suggesting that the hypothetical attractant is not class-specific (Rogulska, Ożdżeński, and Komar 1971).

How many germ cells?

Counts of primordial germ cells, when they can first be identified by staining for alkaline phosphatase activity, average around 50 (Mintz and Russell 1957; Ożdżeński 1967). Over the next few days, during the period of migration and colonization of the genital ridges, the number steadily increases, with a doubling time of a little less than 20 hours (Mintz and Russell 1957; Tam and Snow 1980). The germ cell population in the embryo reaches its maximum size of about 20,000–25,000 by about 14 days *post coitum*. At this time, the female germ cells are beginning to enter meiosis, and the male germ cells undergo mitotic arrest, so little further increase is seen.

When the mother is treated at 7 days *post coitum* with mitomycin C, a potent inhibitor of DNA synthesis that causes massive cell death in the embryo and a temporary suppression of mitotic activity (Snow and Tam 1979), the initial number of germ cells is reduced, along with cell number in the embryo as a whole, to about 15 percent of the control level. At the time the inhibitor is injected, no germ cells can yet be detected. The reduction in their numbers could therefore be due either to the reduction in numbers of an already committed population of germ-cell precursors or to an overall reduction in cell numbers in the epiblast, followed by the selection of a constant proportion to follow the germ-cell pathway. Between 9½ and 10½ days, a markedly increased rate of proliferation is seen in germ cells traversing the dorsal mesentery, leading to a considerable degree of recovery (Snow and Tam, personal communication). However, the numbers are never completely made up, and the mice that develop from these mitomycin-C-treated embryos show a marked deficiency of germ cells in the gonads of both sexes and a high incidence of infertility in males. The relatively uniform cell-cycle time in the control germ-cell population and the marked depression followed by recovery in the treated population are shown in figure 19.

Germ-cell numbers can also be drastically reduced by the cytotoxic drug Busulfan (Merchant 1975). Unlike mitomycin C, Busulfan is most effective when the germ cells are proliferating rapidly within the genital ridges or during the later stages of migration. The somatic cells of the developing gonad seem to be unaffected (Merchant-Larios and Coello 1979). No recovery has been reported, perhaps because at this late stage too little time for proliferation remains, so the mature gonads are almost completely devoid of germ cells.

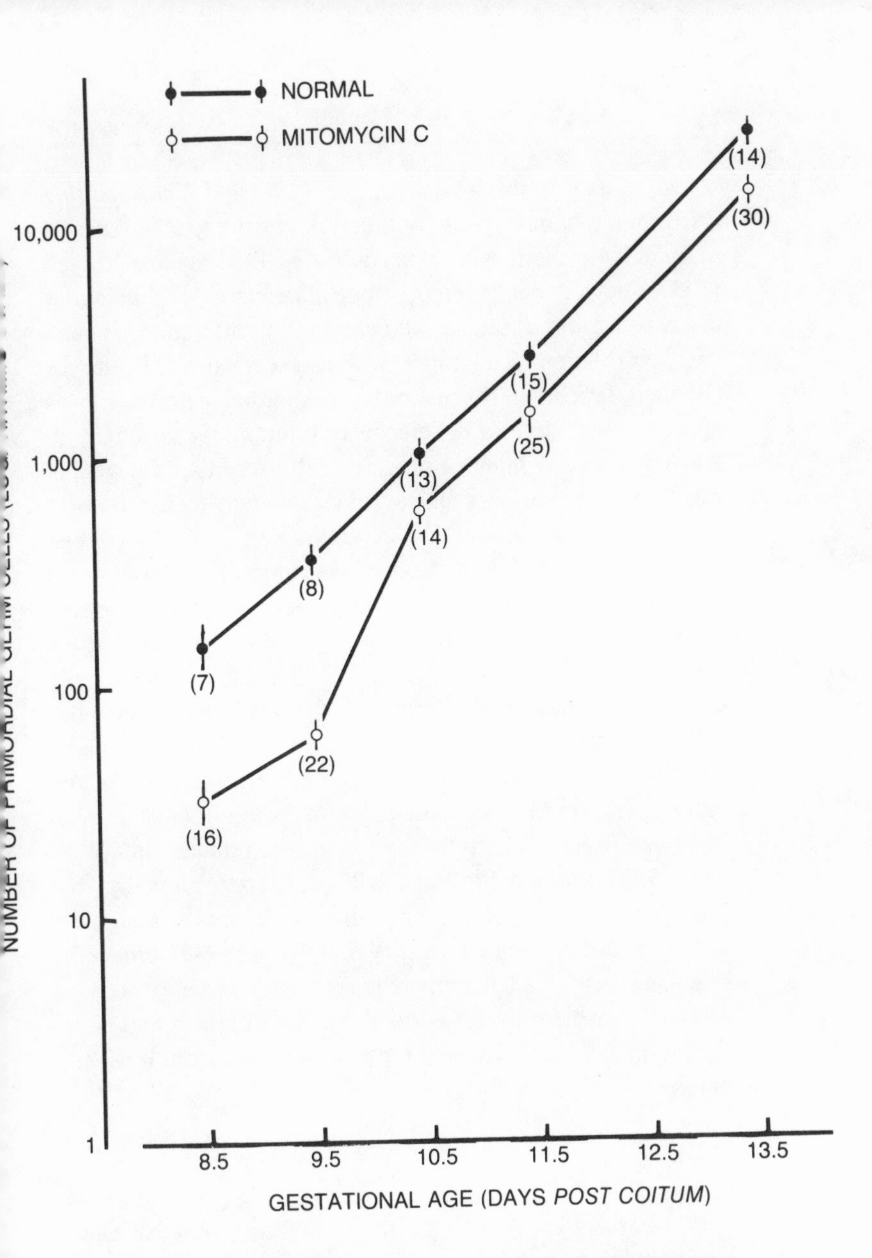

Figure 19. The Number of Primordial Germ Cells, Identified by Their High Alkaline Phosphatase Activity, from 8½ to 13½ Days *post Coitum* in Control Mouse Embryos and in Embryos from Mothers Treated with Mitomycin C at 7 Days *post Coitum*. (From Tam and Snow 1980, figure 5.)

Two mutant genes are known, *Steel* (*Sl*) and *White-spotting* (*W*), that in homozygous condition produce a total absence of germ cells in the newborn or adult gonad. Mintz and Russell (1957) showed that the embryological defect in *W/W* individuals was largely a failure of germ-cell proliferation, though some retardation of migration was also seen. In *Sl/Sl* embryos (Bennett 1956), the formation of primordial germ cells was again normal, and no deficiency in migration was found, so that the reduced number of germ cells reaching the genital ridges must have been entirely due either to their failure to proliferate or to an excessive amount of cell death (McCoshen and McCallion 1975). Although the mode of action of these two genes upon the germ cells appears so similar, a fundamental difference has been revealed by studies on melanocytes (Mayer and Green 1968; Mayer 1973) and hemopoietic cells (McCulloch et al. 1965; Bernstein 1970; McCuskey and Meineke 1973). In *W/W* embryos the defect is cell-autonomous in that the failure of proliferation is an inherent property of the melanocytes and hemopoietic cells themselves, while in *Sl/Sl* embryos, it is a property of the tissue environment. In a *W/W* host, *Sl/Sl* hemopoietic cells will proliferate normally; but *W/W* hemopoietic cells never proliferate normally, regardless of the environment in which they find themselves. Germ cells, like melanocytes and hemopoietic cells, are migratory cells; so probably for them too, it is the tissue environment that is defective in *Sl/Sl* embryos and the cells themselves in *W/W* embryos.

Nuage

Primordial germ cells in the gut wall of the mouse have been shown by electron microscopy to contain aggregates of granulofibrillar material in the cytoplasm (Spiegelman and Bennett 1973). This material was not found in germ cells that had colonized the genital ridges; however, Eddy (1974) described similar material in migrating rat primordial germ cells and claimed that it was also present in most, though not all, subsequent stages of germ-cell development, including oocytes. Both reports agree that the germ-cell cytoplasm at all stages was characterized by a great abundance of ribosomes and polysomes, with relatively little endoplasmic reticulum.

Eddy equated the dense aggregates of granulofibrillar material with nuage, a characteristic component of germ cells in many nonmammalian species. Nuage has since been described in germ cells of mammals, including man (Kellokumpu-Lehtinen and Söderström 1978). Nuage contains protein and sometimes, but not always, RNA; it is often associated with mitochondria and nuclear pores and is thought to resemble the germ plasm of amphibia and some invertebrates (Eddy 1975). Nuage has never been detected in the cytoplasm of preimplantation embryos or epiblast cells, however; it therefore differs from germ plasm in that it cannot be traced continuously from one generation to another. There is also no evidence in mammals that nuage is a germ-cell determinant: germ cells may have nuage because they are germ cells, rather than being germ cells because they have nuage. In *Drosophila*, pole plasm does indeed play a determinative role in germ-cell development (Illmensee and Mahowald, 1974); but even in Amphibia, there is no direct evidence that germ plasm constitutes a germ-cell determinant, and it has been suggested that it may rather serve some function associated with germ-cell migration (Smith and Williams 1979). In spermatogenesis, the nuage finally dissociates from its associated mitochondria at the primary spermatocyte stage, to form the chromatoid body. This body persists until the flagellar midpiece is formed in the spermatid; it then migrates down the axoneme and apparently dissipates (Fawcett, Eddy, and Phillips 1970). The fate of nuage in the mammalian egg is not known.

8 Sexual Differentiation

At the time when the primordial germ cells in the mouse enter and colonize the genital ridges, 10½–11½ days *post coitum*, sexual differentiation has not yet begun. No difference can be detected between male and female embryos, either in the appearance and number of germ cells or in the structure and size of the genital ridges in which they are located. At this stage we start to refer to the genital ridge as an indifferent gonad and to the proliferating germ cells as gonia (oogonia in the female, prospermatogonia in the male).

By 12½ days *post coitum*, the testis in most strains of mice is visibly different from the ovary. It is larger, the blood vessels are larger and arranged differently, and testis cords can for the first time be seen, consisting of solid strings of gonia several cells in diameter, surrounded by a single epithelial layer of somatic cells. Some gonia are left outside the cords, and the cord structure tends to be less well organized near the narrowing stalk of tissue that connects the gonad with the mesonephric rete region of the body wall. The ovary at the same stage appears to some extent divided into compartments by oriented rows of somatic cells, with the germ cells scattered in the stroma between them. In the mouse no cords are seen in the ovary. In some species (e.g., the calf) the gonia are sequestered into cords in both the testis and the ovary, but the testis is distinguishable both by its larger size and the thick *tunica albuginea* that surrounds it.

Gonadal sex

Whether the indifferent gonad differentiates in the male or the female direction depends in the first instance on the sex-chromosome constitution of the embryo and in particular on the presence or absence of a Y chromosome. XY, XYY, and XXY mammals have testes; XO, XX, and XXX mammals have ovaries. But the gonad contains two distinct cell populations, somatic and germinal. Is it the sex chromosome constitution of the somatic tissue or of the germ cells that determines the sex of the gonad?

For a conclusive answer to this question, we would need to study either gonads that had never contained any germ cells or gonads in which the sex-chromosome constitution of the two components differed. As we have seen, the number of germ cells entering the genital ridges can be drastically reduced either genetically (*White-spotting*, *Steel*) or by exposing the embryo to cytotoxic drugs (Busulfan). In all these situations, however, some germ cells reach the genital ridges, and it could be argued that their presence, even in very low numbers, is essential for normal sexual differentiation—in other words, that they exert some triggering effect. In the chick, a recent study involving surgical excision of the anterior germinal crescent from intact embryos (McCarrey and Abbott 1978) succeeded in removing all the primordial germ cells before they reached the genital ridges, while allowing the experimental embryos to continue their development right up to the time of hatching. In both male and female embryos, the morphological and histological differentiation of the somatic elements of the gonad proceeded normally in the complete absence of germ cells. This experiment has not yet proved possible in any mammal.

The other possibility—that of constructing an individual containing a gonad with XX somatic tissue and XY germ cells or vice versa—has also not yet been achieved. Studies on mouse and human chimeras and mosaics, in which varying proportions of XX and XY or XO and XY somatic and/or germinal cells coexist in the gonads, have so far proved inconclusive (see McLaren 1976c). However, the weight of evidence from all sources favors the view that the initial direction of sexual differentiation in the mammalian gonad is dictated by the somatic tissue rather than by the germ cells. Subsequent development of the gonad may of course be strongly influenced by the presence or absence of germ cells: for example, if the oocytes in an ovary all degenerate, then no follicles are maintained, no estrogen or progesterone is produced, and a "streak" gonad results, as seen in XO women with Turner's syndrome.

There is increasing evidence that the Y chromosome exerts its effect on sexual differentiation through the male histocompatibility antigen, H-Y antigen (Wachtel, Ohno, Koo, and Boyse 1975). The presence of this cell-surface antigen is so strongly correlated with the presence of a Y chromosome that either it must be coded for by a Y-chromosome gene or, if the structural gene is located on some other chromosome, the Y chromosome must carry a regulator gene (for discussion, see Wachtel and Koo 1980). To explain the observation that embryonic gonads (for example, in mouse chimeras) containing a mixture of XX and XY cells tend to differentiate as normal testes rather than as ovaries or ovotestes, it is postulated by Ohno that H-Y antigen can be released from XY cells and become attached to receptors on the surface of XX cells, where it exerts an organizing role in the direction of testicular differentiation. Attempts to confirm this hypothesis experimentally have aimed either to induce testicular differentiation by exposing XX gonads to H-Y antigen *in vitro* (Zenzes et al. 1978) or to prevent such differentiation by removing H-Y antigen from XY cells (Ohno, Nagai, and Ciccarese 1978). Although the results of these experiments have so far been suggestive rather than conclusive, the basic hypothesis that H-Y antigen in some way induces primary testicular differentiation remains highly plausible.

Germ-cell sex

Meanwhile, what of the germ cells? We left them steadily increasing in number up to about 14 days' gestation. On the previous day, as we now know, sexual differentiation of the gonad became apparent, so that the male germ cells—the prospermatogonia—found themselves packed together inside testicular cords. Now the germ cells as well start to undergo sexual differentiation. The oogonia, after a final mitotic division and a final round of DNA replication, enter the prophase of meiosis and pass, in a couple of days, through leptotene and zygotene into pachytene. In the male, the prospermatogonia cease dividing at about the time that the oogonia are entering meiosis and become arrested in the G1 stage of the cell cycle. The resumption of mitotic proliferation after birth and the transition from prospermatogonia to spermatogonia have been well documented by Hilscher and Hilscher (1976). Male germ cells finally enter the prophase of meiosis 9–10 days after birth.

We may ask the same questions about germ-cell differentiation that we asked about gonadal differentiation. Is it the germ cells themselves or the somatic tissues that determine the direction of germ-cell differentiation? What role do the sex chromosomes play? What about H-Y antigen?

Experimental mouse chimeras made by aggregating together two 8-cell embryos, one male and one female, normally develop as phenotypic males, even though they contain a population of XX cells as well as a population of XY cells. The testis of such an individual would be expected to contain XX as well as XY germ cells, and the fate of the XX germ cells should tell us something about the cause of germ-cell differentiation.

In Amphibia, female germ cells developing in a testis give rise to functional spermatozoa, but breeding experiments on XX/XY chimeras have given ample evidence that the same is not true for mice. XX germ cells do not even contribute to the spermatocyte population, as a large number of chromosome preparations made from the testes of XX/XY chimeras have yielded only XY meiotic plates (Mystkowska and Tarkowski 1968; McLaren 1975b). The first clue as to the fate of the XX germ cells came from the report of Mystkowska and Tarkowski (1970), that fetal XX/XY testes contained some germ cells in meiotic prophase, and we have confirmed this finding (figure 20). Now, since germ cells entering meiosis before birth are normally confined to the ovary, as we have seen, and since male germ cells normally do not enter meiosis until 9–10 days after birth, it was tempting to assume that these prenatal meiotic cells were in fact the XX germ cells behaving autonomously and entering meiosis according to their own developmental program. There was, however, another possibility. Since the chimeric testis contained two populations of somatic cells as well as two populations of germ cells, it could be that

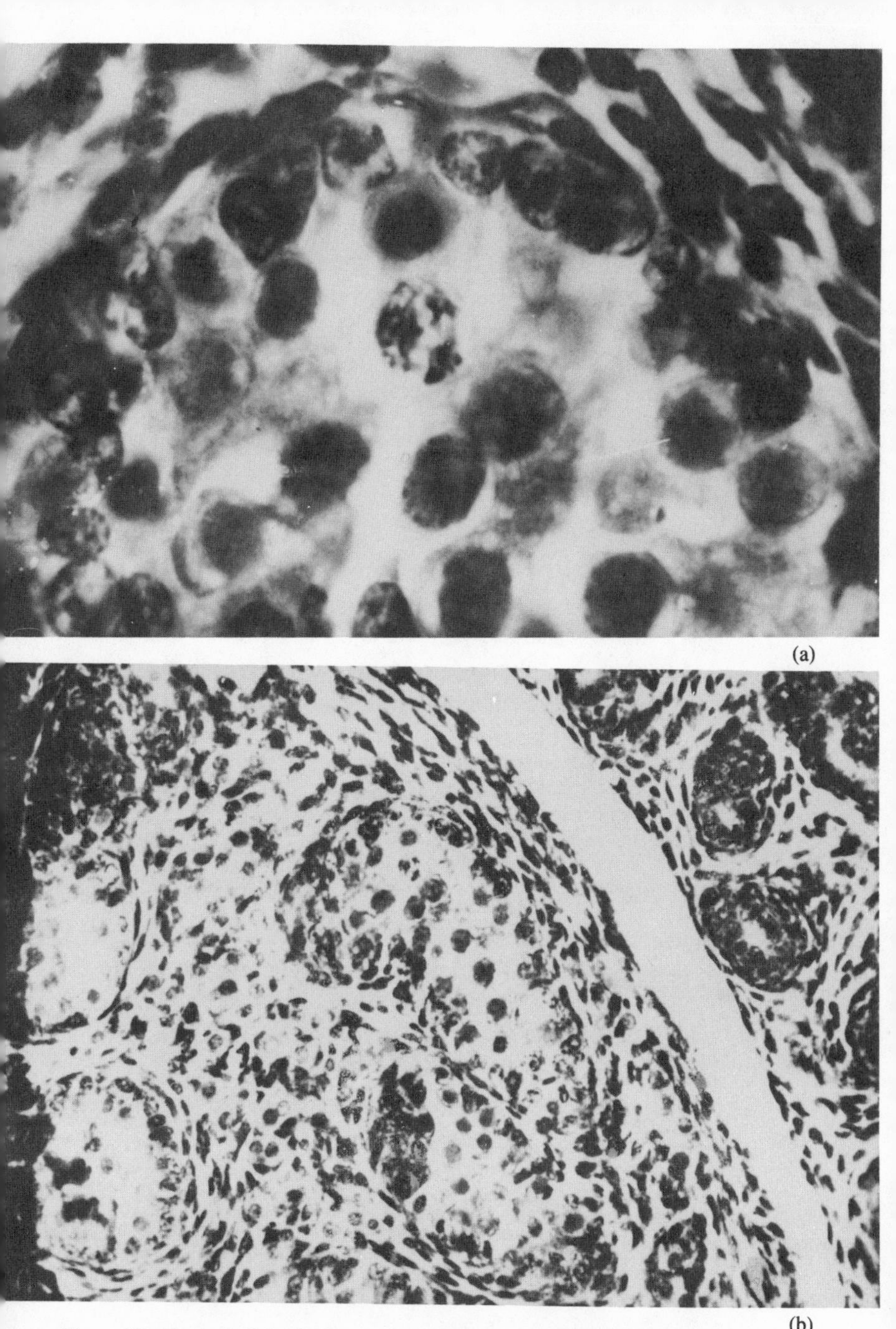

Figure 20. Germ Cell in the Prophase of Meiosis in the Fetal Testis of a Mouse Aggregation Chimera 16½ Days *post Coitum*. (a) High-power view, showing one germ cell in the pachytene stage of prophase among several prospermatogonia in interphase. (b) Low-power view, showing the position of the testis cord in which the meiotic germ cell is located, in the vicinity of the mesonephric rete region.

the XX somatic tissue was somehow inducing prenatal meiosis in any germ cell in its neighborhood, be it XX or XY. In normal development, male meiotic germ cells are distinguished from female by the precociously condensing XY bivalent, located in a sex vesicle that labels strongly with [^{3}H]thymidine. We looked for such a vesicle in the meiotic germ cells of the fetal chimeric testis (McLaren, Chandley, and Kofman-Alfaro 1972) but failed to find it.

This study provided circumstantial evidence that it was indeed the XX germ cells that were entering meiosis before birth, at just the same time as they would have in an ovary. Some problems remained, however. First, the evidence was indirect, and it was possible that the absence of a sex vesicle was a characteristic of prenatal entry into meiosis rather than a reflection of chromosome constitution. Second, there were very few meiotic cells in the chimeric testes before birth, certainly nothing like the 50 percent that one would expect on average for the XX component of an XX/XY chimera.

Meiosis-Inducing Substance

A further problem emerged when Byskov (1974) demonstrated that entry into meiosis was not an autonomous property of germ cells but was induced by a substance diffusing into the gonad from the adjacent mesonephric rete region. *In vitro* experiments showed that female germ cells entered meiosis if the indifferent gonad was cultured with the rete attached, but not if the gonad was isolated (Byskov and Saxen 1976). Later work suggested that the hypothetical meiosis-inducing substance (MIS) was produced by both the female and the male rete (Byskov 1978), but not until 12–13 days *post coitum*. By this time, the male germ cells would normally be sequestered in testis cords, where they would be protected from the action of MIS, perhaps by a still more hypothetical meiosis-preventing substance (MPS) secreted by the pre-Sertoli cells lining the cords. The secretion of MPS may eventually cease, so that in the postnatal period the male germ cells too are able to respond to MIS and enter meiosis (Grinsted, Byskov, and Andreasen 1979). However, if male gonads were removed early enough, before testis cords had formed, and were cultured together with female retes taken at a stage when they were already producing MIS, the male germ cells could be induced to enter meiosis precociously (Byskov and Saxen 1976). Similar conclusions were reached by O & Baker (1976, 1978) for the hamster; they cultured intact gonads, with and without the rete, and artificial mixtures of XX and XY somatic and germ cells. The ability of male germ cells to enter meiosis precociously when the testis cord structure was disorganized had been demonstrated previously by Ożdżeński (1972), in mouse genital ridges transplanted beneath the adult kidney capsule. Those male germ cells that fail to be incorporated into testis cords may also enter meiosis before birth if they are located in the vicinity of the mesonephric rete, but not otherwise (Byskov 1978).

Now, if male and female germ cells are equally capable of responding to MIS, given that they are equally exposed to it, what becomes of our evidence that XX, but not XY, germ cells entered meiosis before birth in XX/XY chimeric testes? The chimera situation is complicated by the presence of two populations of germ cells, and so I decided to look at a different situation, one where all the germ cells in a testis are XX in chromosome constitution. This occurs in mice carrying the dominant gene *Sex-reversed* (*Sxr*), which converts an XX embryo into a phenotypic male (Cattanach, Pollard, and Hawkes 1971).

Cattanach and his colleagues reported that the XX germ cells in the *Sxr*/ + testis were capable of differentiating into Type A, and even Type B, spermatogònia but degenerated soon after birth. I examined the testes at an earlier stage, when XX germ cells in an ovary would be entering meiosis, and found that the XX *Sxr*/+ testes, as in the earlier case of the XX/XY chimeras, contained some germ cells in the prophase of meiosis (figures 21a and b). They were few in number, as in the chimeras, and were all located in the immediate vicinity of the mesonephric rete (McLaren, 1981).

Thus, the XX germ cells appear to be responding to MIS even though they are inside cords. But why should XX germ cells react in this way, and not XY? It seems that the chromosome constitution of a germ cell influences its susceptibility to the meiosis-inducing influence of MIS. Outside cords, XX germ cells in the ovary almost always enter meiosis before birth, while XY germ cells in the testis sometimes but not always do so. XY germ cells in an ovary may also occasionally enter meiosis before birth, since there is one report of an XY germ cell in the ovary of a mouse chimera undergoing oogenesis (Evans, Ford, and Lyon 1977). Inside cords the evidence from *Sxr*/+ and XX/XY chimeras suggests that XX germ cells sometimes, but not always, enter meiosis before birth, while XY germ cells hardly ever do so.

The chromosome constitution of XX and XY cells differs in two respects: the number of X chromosomes and the presence or absence of a Y chromosome. XO germ cells should provide a crucial test of which factor affects responsiveness to MIS, since they resemble germ cells in having only one X chromosome and XX germ cells in lacking a Y chromosome. In the postnatal *Sxr*/+ testis, XO differ from XX germ cells in that they are capable of undergoing spermatogenesis, though the spermatozoa that are produced are abnormal and incapable of fertilization (Cattanach, Pollard, and Hawkes 1971). An ex-

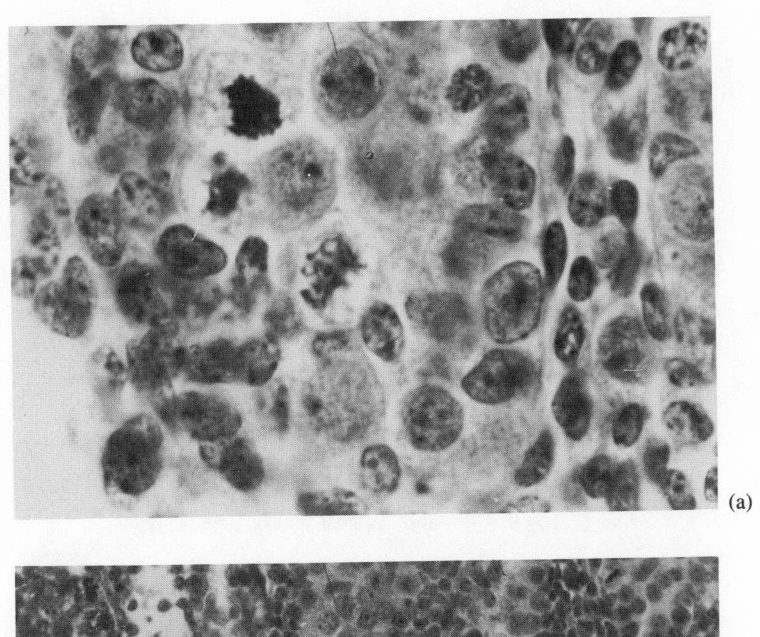

(a)

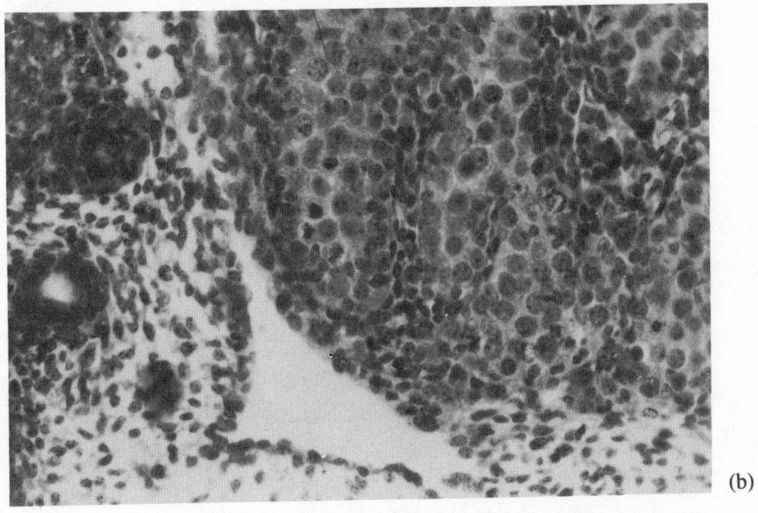

(b)

Figure 21. Germ Cells in the Prophase of Meiosis in the Fetal Testis of an XX *Sxr*/+ Mouse 16 Days *post Coitum*. (a) High-power view, showing some germ cells in the pachytene stage of prophase and some prospermatogonia in interphase. (b) Low-power view, showing the position of the testis cord in which the meiotic germ cells are located, in the vicinity of the mesonephric rete region.

Table 5. Presence of Meiotic Germ Cells in 15- to 16-Day Fetuses from XO Females Mated to XY *Sxr*/+ Males

Genotype of fetus	Phenotypic sex	No. of gonads: No. with meiotic germ cells	No. of gonads: Total
XX, +/+	♀	24	24
XO, +/+	♀	6	6
XY, +/+ or *Sxr*/+	♂	2	55
XX, *Sxr*/+	♂	22	29
XO, *Sxr*/+	♂	2	18

Source: McLaren 1980 and unpublished observations.

amination of fetal XO *Sxr*/+ testes has shown (table 5) that they too differ from XX testes in that their germ cells very rarely enter meiosis precociously, even those close to the rete (McLaren 1981). Preliminary evidence (McLaren and Burgoyne; unpublished observations) on the fate of XO germ cells in the fetal ovary suggests that they are less responsive to MIS there also. If these findings are confirmed, it would show that XO resemble XY germ cells in being less responsive to MIS in the embryo than XX germ cells, both when they are sequestered inside cords and when they are not (table 6). It appears therefore to be the presence of a second X chromosome that facilitates entry into meiosis, and the presence of the Y chromosome is from this point of view irrelevant.

These observations clarify the earlier chimera findings in that they support the view that it is the XX germ cells that enter meiosis before birth and explain why so few meiotic cells are observed. Although the stimulus to enter meiosis comes from the environment, the germ cell's response depends both on its location and on its own chromosome constitution.

Table 6. Meiotic Status of Fetal Germ Cells in Relation to Their Location and Sex-Chromosome Constitution

Location	Sex chromosome constitution	Germ cells		Source
		In meiosis	Not in meiosis	
Inside testis cords	XX	+	+	*Sxr*/+ XX[a]
	XO	−	+	*Sxr*/+ XO[a]
	XY	−	+	Normal or *Sxr*/+ XY testis
Not inside testis cords	XX	+	−	Normal XX ovary
	XO	+	+?	XO ovary[b]
	XY	+	+	Germ cells excluded from cords in normal testis[c]

[a]McLaren 1980.
[b]Preliminary observations by McLaren and Burgoyne.
[c]Byskov 1978.
Source: McLaren 1980.

X-chromosome activation

The hypothesis that embryonic germ cells with two X chromosomes are more responsive to MIS than those with only one implies that both X chromosomes must be functioning at the time when the embryonic germ cells are exposed to MIS—that is, immediately before they are observed to enter meiosis. We have already seen that inactivation of the second X chromosome in epiblast cells takes place shortly before gastrulation, while during oogenesis both X chromosomes are active. Figure 22 shows that the gap between gastrulation and oogenesis could in theory be filled in two different ways. Either the germ-cell precursors could undergo X-inactivation along with the rest of the epiblast, which would mean that at some point reactivation would have to occur (figure 22a); or the germ line could escape inactivation completely (figure 22b). What is known about X-chromosome activity in germ cells during this crucial interval?

In human embryos heterozygous for G6PD, Gartler, Andina, and Gant (1975) were able to detect hybrid dimer, indicating that both X chromosomes were active, in ovaries containing some meiotic cells, but not in those where only oogonia were present. They therefore suggested that meiosis was the trigger for X-chromosome reactivation. Migeon and Jelalian (1977), on the other hand, found evidence of hybrid dimer in ovaries as early as 8 weeks of fetal age, well before the onset of meiosis, confirming that both X chromosomes are functioning before meiosis begins. The authors favor the view that female germ cells may escape X-inactivation altogether.

In the mouse, Andina (1978) examined the activity of two X-coded enzymes, G6PD and HPRT, in oocytes from XX and XO mice. He found a twofold activity difference, indicating that both X chromosomes were functional in the XX germ cells in meiotic stages, but he did not examine any premeiotic germ cells. We have therefore looked at germ cells at earlier times, using the ratio of activity of the X-coded HPRT to the autosomal adenine phosphoribosyl transferase (APRT) as an index of X-chromosome function (Monk & McLaren 1981). Once the XX germ cells have entered meiosis, the HPRT:APRT ratio rises dramatically; some of this increase appears to be due to biochemical events associated with meiosis as such. At 12½ days, even before the onset of meiosis, however, the XX germ cells show a significant increase, while at 11½ days, immediately after the germ cells have colonized the genital ridges, we could detect no difference between the ratio in XX, XO, and XY germ cells.

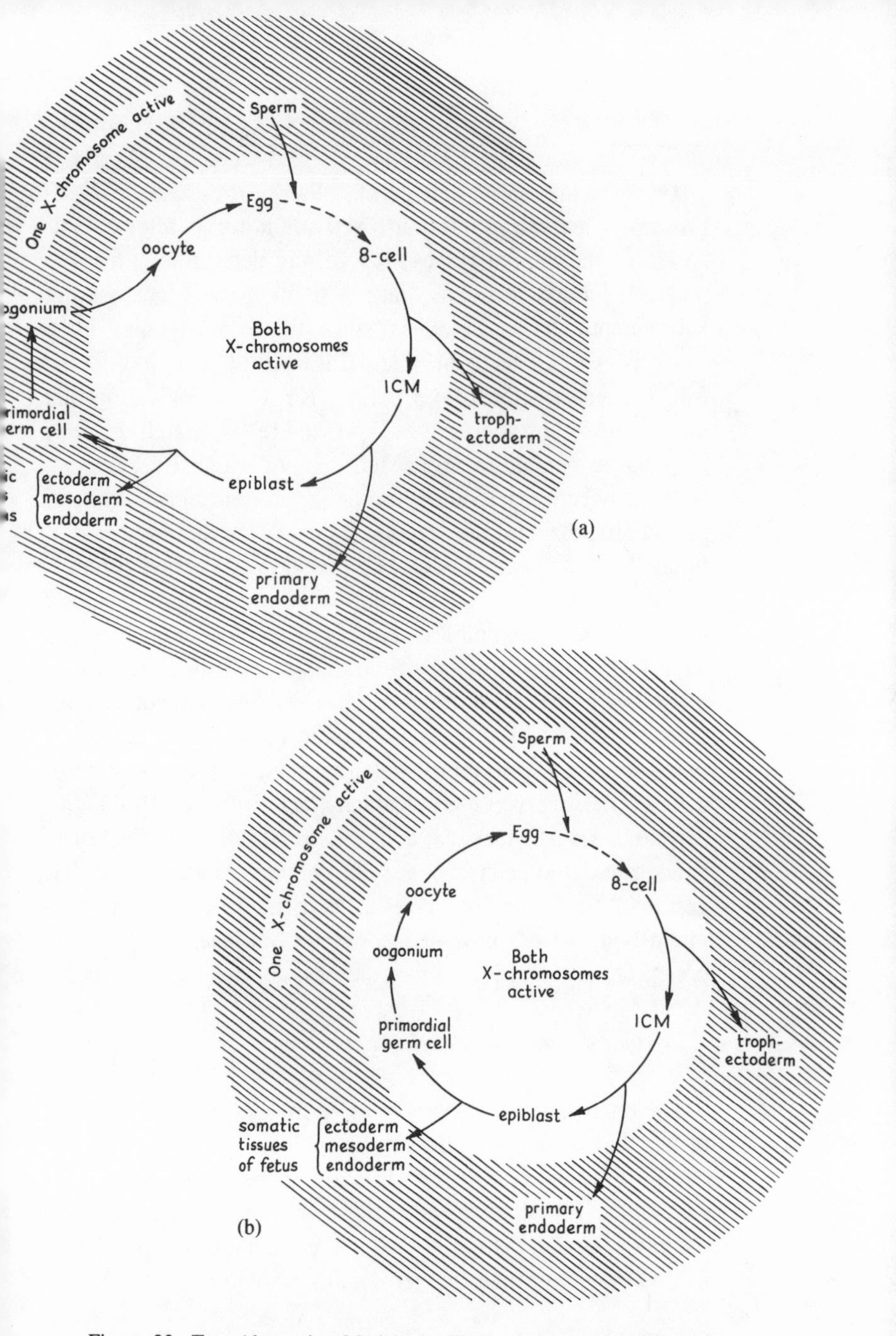

Figure 22. Two Alternative Models for X-Chromosome Activity of Germ Cells during the Period between Gastrulation and Oogenesis. (After Monk and Harper 1979, figure 1.) (a) Germ-cell precursors in XX embryos have one X chromosome inactivated along with the rest of the epiblast, followed by subsequent reactivation; (b) the germ-cell line in XX embryos never undergoes X-chromosome inactivation.

These results show that germ cells do not escape X-chromosome inactivation but that reactivation occurs after colonization of the genital ridges, before any detectable entry into meiosis. Indeed, it seems that reactivation can occur without subsequent entry into meiosis, since prospermatogonia from a *Sex-reversed* XX testis have an HPRT:APRT ratio double that found in germ cells from a normal XY testis, indicating that the second X chromosome is expressed, even though the cells have not responded to MIS (McLaren and Monk 1981).

As to when inactivation of the X chromosome occurs in germ cells, the simplest hypothesis would be that indicated in figure 22a, namely that germ-cell precursors, whether committed or not, undergo X-inactivation before gastrulation, along with the rest of the epiblast. However, no biochemical studies have been carried out on primordial germ cells during their migratory period, and the cytological findings are conflicting. Ohno claimed to see a sex chromatin body in migrating primordial germ cells (Ohno 1964), but none in premeiotic germ cells of the ovary (Ohno, Kaplan, and Kinosita, 1961). Semyonova-Tian-Shanskaya and Patkin (1978), on the other hand, could find no sign of sex chromatin in human primordial germ cells until they had entered the genital ridges. Sex chromatin bodies were present in oogonia but disappeared at the preleptotene stage, just before the onset of meiosis.

Postnatal germ-cell development

The embryonic development of germ cells, and the question of whether or not they embark on meiosis, constitutes one of the most crucial points of interaction of the germ cell with its somatic environment and illustrates well the interplay between the environmental stimulus and the autonomous chromosomal factors that determine the germ cell's response. We must now turn our attention to the postnatal consequences of these prenatal events.

In the normal mouse testis, the first spermatogonia to enter meiosis do so about 9–10 days after birth, and successive waves continue to enter meiosis at intervals throughout life, leaving a continuously proliferating stem-cell population. The Y chromosome is said, on cytological grounds, to be active in later spermatogonia (Ghosal and Mukherjee 1977) but to be inactivated along with the X chromosome on entry into meiosis (Lifschytz and Lindsley 1972). No RNA synthesis in the condensed XY bivalent could be detected at any stage of meiosis in male germ cells of the mouse, though the autosomes during the same period exhibit a well-defined pattern of RNA synthesis (Monesi 1971). XO germ cells in the *Sex-reversed* testis appear normal throughout the spermatogonial and early spermatocyte stages of spermatogenesis, but many degenerate at later stages, and the few spermatozoa that are formed have abnormal heads and show little or no motility (Cattanach, Pollard, and Hawkes 1971). It is not known whether these abnormalities are due to absence of the Y chromosome or to the *Sxr* gene.

XX germ cells in the *Sex-reversed* testis, as we have seen, can take either the female or the male developmental pathway, depending, it seems, on whether or not they are acted on by MIS before birth. If they take the male pathway, they all degenerate within a few days of birth, as spermatogonia (Cattanach, Pollard, and Hawkes 1971). It appears that two X chromosomes are incompatible with spermatogenesis (see Lyon 1974; Burgoyne 1978), and reports that female germ cells cultured in the vicinity of a testis embark on spermatogenesis (Turner 1969) have not been confirmed (Ożdżeński et al. 1976). If the germ cells take the female pathway and enter meiosis in the embryo, most degenerate before birth—probably during the wave of atresia that hits female germ cells during pachytene—but a few survive birth and are found as growing oocytes (see figure 4) in the seminiferous tubules 1–2 weeks after birth (McLaren 1980). None survives into the adult testis, perhaps because the follicle surrounding the oocyte fails to develop. Very much the same picture—of growing oocytes within the seminiferous tubules—is seen when XY germ cells are induced to enter meiosis precociously by transplantation of the genital ridge under the kidney capsule, with consequent disruption of the cord structure (Ożdżeński 1972). So it seems that a Y chromosome does not prevent oogenesis if the germ cell has once embarked on the female pathway of development by entering meiosis before birth.

Indeed, in the ovary, not only has an XY oocyte been detected in an XX/XY chimera (Evans, Ford, and Lyon 1977), but an egg containing a Y chromosome was actually shed by such a chimera and was fertilized and developed normally (Ford et al. 1975). However, this case was complicated by the fact that the resulting individual proved to be XXY in chromosome constitution. If, as seems likely, a normal X-bearing spermatozoon had fertilized an XY egg, the egg may have developed from an XXY germ cell that arose by nondisjunction from an XY precursor. The presence of a second X chromosome may therefore have contributed to its successful oogenesis. Germ cells with only one X chromosome seem to be at a considerable disadvantage during oogenesis, when both X chromosomes are normally switched on. XO oocytes degenerate around the time of birth in man, and XO female mice, though fertile, have a shorter reproductive life span than do normal XX females. Burgoyne (1978) has suggested that the lack of a second X chromosome during oogenesis becomes progressively more of a handicap with time, so that XO oocytes only survive into sexual maturity in the mouse because of its shorter life cycle. As we have seen, Burgoyne and Biggers (1976) found a high rate of embryonic loss in XO mothers during the first few cleavage divisions, suggesting that the eggs were indeed abnormal. XO ovaries contained fewer oocytes than XX ovaries at all the ages examined (Burgoyne and Baker 1980), perhaps because a smaller proportion of the germ cells responded to MIS and entered meiosis before birth.

In the normal mouse ovary, waves of atresia, both before and after birth, reduce the oocyte population substantially. In the adult female, the rate of atresia is, at least to some extent, under hormonal control, since it decreases after hypophysectomy (Jones and Krohn 1961). Whether the same is true for the infant ovary, and whether the properties of the oocyte itself determine to any extent its chances of survival, or whether the impact of atresia is entirely random, we have no idea.

Follicle cells are already beginning to surround the oocytes before birth in the mouse. By about 5 days after birth, the nucleus is in the dictyate condition and once more in meiotic arrest, the primordial follicles are fully formed, and the first oocytes are about to leave the pool and embark on their period of growth.

9
Conjectures

We have now taken our mouse germ cell through its entire life cycle, from oocyte growth and maturation through ovulation, fertilization and the development of the fertilized egg, through cleavage and implantation up to gastrulation; through the reappearance of germ cells, their migration, colonization, and sexual differentiation; up to the formation of the follicle, from which the growth phase springs once again.

It is clear that the differentiated germ cell exists in most intimate association and constant interaction with the somatic tissues that constitute its environment. Its migration is guided along an ordained route; its entry into meiosis and the direction of differentiation depend on signals from the environment; once in a follicle, the female germ cell is nourished and controlled by the surrounding follicle cells; during oocyte growth, proteins are pumped into it in large amounts; its maturation is regulated by a precise balance of hormones. In return, the female germ cell provides a gamete that has undergone meiosis, with a surface specialized for fusion with the spermatozoon and cytoplasm specialized for decondensing its nucleus. The gamete is programmed for cleavage and compaction and is provided with a large store of mitochondria, ribosomes, proteins (including histones), and RNA (including messenger RNA) to provide for the embryo during at least the first two days of its life, until its own supply system can take over.

We are still almost totally ignorant of most aspects of the interaction between germ cells and soma. For example, we have no idea how the primordial germ cells are guided to the genital ridges; whether the somatic environment is necessary for the survival of germ cells at this stage, or whether they could be isolated and grown for an indefinite period in culture; how the X chromosome is switched on; how the signal for meiosis is transmitted and received; how meiotic arrest is achieved; what determines whether and when germ cells undergo atresia and degenerate; what determines when they leave the primordial pool and start to grow; what degenerative processes occur during the long period of arrest; what proteins are taken up from the follicle; or how much of the protein either taken up or synthesized is important for embryonic development.

Then there is the whole problem of a possible germ-cell lineage in the pregastrulation embryo and the salvage of totipotency. It seems that some restrictions of developmental potential must have occurred in the ancestry of the germ cells. The late inner cell mass is no longer able to regenerate trophectoderm derivatives, and the late epiblast can no longer regenerate primary endoderm, so it is hard to see how a truly totipotent cell line could be maintained throughout embryogenesis, except in the sense that any cell line that contributes to a germ-cell population is ultimately totipotent. Changes in the pattern of transcription occur during differentiation; these may be irreversible, but subsequent differentiation may bring about a series of further changes that eventually recreate the original pattern. In the carrot, every somatic cell is capable of giving rise to an entire new carrot plant, including germ cells. We do not think the same is true for a mouse, but perhaps we are merely ignorant of the right environmental cues to bring about the requisite changes. Perhaps, at least, any epiblast cell can give rise to germ cells.

We do not know what is special about germ cells. We do not know whether nuage is a germ-cell determinant, and whether it will be found in pregastrulation embryos, and if so, whether it will be present in all cells or only a subset of cells. Nor do we know the relation between the pluripotent embryonal carcinoma stem cells and the primordial germ cells or epiblast cells or a subset of the epiblast cells. I would like to end by making three conjectures.

Conjecture 1. That the potential for totipotency resides in all cells. This would be contradicted if, for example, a nuagelike germ-cell determinant were discovered that could in no way be generated *de novo* within a cell, but could only be transmitted from one cell to another (the germ-plasm concept in its purest form).

Conjecture 2. That to realize this potential for totipotency, the cell must be exposed to the appropriate environment. In normal development, this environment is created by the somatic tissues in association with which the germ cells develop. We do not know at what point in the differentiation of a germ cell it achieves totipotency: it could be early, or it could be very late.

Conjecture 3. That the selection of certain cells to form the germ line and others to form the somatic tissues depends on their position within the epiblast. This is the most readily testable of the three conjectures and hence the most interesting.

The problem of the relationship between germ cells and soma with which we began has resolved itself into two separate problems: the extent to which germ cells are susceptible to somatic influence and the origin of germ cells. With regard to the first problem, we have seen in chapter 2 that although the mammal is arranged in such a way that the developing germ cells are relatively sequestered and protected from environmental insults, the germ cells are by no means wholly immune from the influence of hormones and other substances emanating from the soma. Indeed, in chapter 7 it emerged that such substances may affect not only whether a germ cell survives or perishes but even the direction in which it differentiates, male or female. Clearly the Weismannist view set out in figure 1a is not tenable. It is to the second problem, of the origin of the germ cells, that my conjectures relate. If they were demolished by future research, then some compromise between figure 1a and figure 2 might constitute an acceptable model. But if they were upheld, even in part, then figure 2 could be regarded as a valid updating of figure 1b, since an absolute distinction between germ cells and soma could no longer be made.

References

Adler, D. A., West, J. D., and Chapman, V. M. (1977) Expression of α-galactosidase in preimplantation mouse embryos. *Nature 267*, 838–39.

Anderson, E., and Albertini, D. F. (1976) Gap junctions between the oocyte and companion follicle cells in the mammalian ovary. *J. Cell Biol., 71*, 680–86.

Andina, R. J. (1978) A study of X chromosome regulation during oogenesis in the mouse. *Exp. Cell Res. 111*, 211–18.

Bachvarova, R., and De Leon, V. (1980) Polyadenylated RNA of mouse ova and loss of maternal RNA in early development. *Dev. Biol. 74*, 1–8.

Baker, T. G., Challoner, S., and Burgoyne, P. S. (1980) The number of oocytes and the rate of atresia in hemiovariectomized mice up to eight months after surgery. *J. Reprod. Fert. 60*, 449–56.

Baker, T. G., and Neal, P. (1973) Initiation and control of meiosis and follicular growth in ovaries of the mouse. *Annls Biol. anim. Biochim. Biophys. 13*, 137–44.

Bedford, J. M., and Cross, N. L. (1978) Normal penetration of rabbit spermatozoa through a trypsin- and acrosin-resistant zona pellucida. *J. Reprod. Fert. 54*, 385–92.

Bennett, D. (1956) Developmental analysis of a mutant with pleiotropic effects in the mouse. *J. Morph. 98*, 199–234.

Bernstein, S. E. (1970) Tissue transplantation as an analytic and therapeutic tool in hereditary anemias. *Amer. J. Surg. 119*, 448–51.

Berrios, M., and Bedford, J. M. (1979) Oocyte maturation: aberrant post-fusion responses of the rabbit primary oocyte to penetrating spermatozoa. *J. Cell Sci. 39*, 1–12.

Biggers, J. D., and Borland, R. M. (1976) Physiological aspects of growth and development of the preimplantation mammalian embryo. *A. Rev. Physiol. 38*, 95–119.

Blandau, R. J., White, B. J., and Rumery, R. E. (1963) Observations on the movements of the living primordial germ cells in the mouse. *Fert. Steril. 14*, 482–89.

Braude, P. R. (1979) Time-dependent effects of α-amanitin on blastocyst formation in the mouse. *J. Embryol. exp. Morph. 52*, 193–202.

Braude, P., Pelham, H., Flach, G., and Lobatto, R. (1979) Post-transcriptional control in the early mouse embryo. *Nature 282*, 102–05.

Brinster, R. L., Chan, H. Y., Trumbauer, M. E., and Avarbock, M. R. (1980) Translation of globin messenger RNA by the mouse ovum. *Nature 283*, 499–501.

Brown, S. W., and Chandra, H. S. (1973) Inactivation system of the mammalian X chromosome. *Proc. Nat. Acad. Sci. U.S.A. 70*, 195–99.

Buehr, M., and McLaren, A. (1974) Size regulation in chimaeric mouse embryos. *J. Embryol. exp. Morph. 31*, 229–34.

Burgoyne, P. S. (1975) Sperm phenotype and its relationship to somatic and germ line genotype: a study using mouse aggregation chimeras. *Dev. Biol. 44*, 63–76.

———. (1978) The role of the sex chromosomes in mammalian germ cell differentiation. *Annls Biol. anim. Biochim. Biophys. 18*, 317–25.

Burgoyne, P. S., and Baker, T. G. (1981) A study of oocyte depletion in XO mice and their XX sibs from 12–200 days *post partum*. *J. Reprod. Fert. 61*, 207–12.

Burgoyne, P. S., and Biggers, J. D. (1976) The consequences of X-dosage deficiency in the germ line: impaired development *in vitro* of pre-implantation embryos from XO mice. *Dev. Biol. 51*, 109–17.

Byskov, A. G. (1974) Does the rete ovarii act as a trigger for the onset of meiosis? *Nature 252*, 396–97.

———. (1978) The meiosis inducing interaction between germ cells and rete cells in the fetal mouse gonad. *Annls Biol. anim. Biochim. Biophys. 18*, 327–34.

Byskov, A. G., and Saxen, L. (1976) Induction of meiosis in fetal mouse testis *in vitro*. *Dev. Biol. 52*, 193–200.

Calarco, P. G., and Brown, E. A. (1969) An ultrastructural and cytological study of the preimplantation development in the mouse. *J. exp. Zool. 171*, 253–84.

Canipari, R., Pietrolucci, A., and Mangia, F. (1979) Increase of total protein synthesis during mouse oocyte growth. *J. Reprod. Fert. 57*, 405–13.

Cattanach, B. M., Pollard, C. E., and Hawkes, S. G. (1971) Sex-reversed mice: XX and XO males. *Cytogenetics 10*, 318–37.

Chiquoine, A. D. (1954) The identification, origin and migration of the primordial germ cells in the mouse embryo. *Anat. Rec. 118*, 135–46.

Clark, J. M., and Eddy, E. M. (1975) Fine structural observations on the origin and associations of primordial germ cells of the mouse. *Dev. Biol. 47*, 136–55.

Cooper, D. W., Johnston, P. G., Murtagh, C. E., Sharman, G. B., Vandeburg, J. L., and Poole, W. E. (1975) Sex-linked isozymes and sex chromosome evolution and inactivation in kangaroos. In *Isozymes*, Vol. 3, *Developmental biology*, ed. L. C. Markert, pp. 559–73. New York: Academic Press.

Copp, A. J. (1978) Interaction between inner cell mass and trophectoderm of the mouse blastocyst. I. A study of cellular proliferation. *J. Embryol. exp. Morph. 48*, 109–25.

———. (1979) Interaction between inner cell mass and trophectoderm of the mouse blastocyst. II. The fate of the polar trophectoderm. *J. Embryol. exp. Morph. 51*, 109–20.

Cross, P. C., and Brinster, R. L. (1970) *In vitro* development of mouse oocytes. *Biol. Reprod. 3*, 298–307.

Darlington, C. E. (1953) *The facts of life*. London: Allen and Unwin.

Dekel, N., and Beers, W. H. (1978) Rat oocyte maturation in vitro: relief of cyclic AMP inhibition by gonadotrophins. *Proc. Nat. Acad. Sci. U.S.A. 75*, 4369–73.

Dubois, R. (1968) La colonisation des ébauches gonadiques par les cellules germinales de l'embryon de Poulet, en culture *in vitro*. *J. Embryol. exp. Morph. 20*, 189–213.

Dubois, R., and Croisille, Y. (1970) Germ-cell line and sexual differentiation in birds. *Phil. Trans. Roy. Soc. Lond. B. 259*, 73–89.

Ducibella, T., Albertini, D. F., Anderson, E., and Biggers, J. D. (1975) The preimplantation mammalian embryo: characterisation of intercellular junctions and their appearance during development. *Dev. Biol. 45*, 231–50.

Ducibella, T., Ukena, T., Karnovsky, M. and Anderson, E. (1977) Changes in cell shape and cortical cytoplasmic organisation during embryogenesis of the preimplantation mouse embryo. *J. Cell Biol. 74*, 153–67.

Dziadek, M. (1979) Cell differentiation in isolated inner cell masses of mouse blastocysts *in vitro*: onset of specific gene expression. *J. Embryol. exp. Morph. 53*, 367–79.

Eager, D. D., Johnson, M. H., and Thurley, K. W. (1976) Ultrastructural studies on the surface membrane of the mouse egg. *J. Cell Sci. 22*, 345–53.

Eddy, E. M. (1974) Fine structural observations on the form and distribution of nuage in germ cells of the rat. *Anat. Rec. 178*, 731–58.

———. (1975) Germ plasm and the differentiation of the germ cell line. *Int. Rev. Cytol. 43*, 229–80.

Epel, D. (1978) Mechanism of activation of sperm and egg during fertilization of sea urchin gametes. In *Current topics in developmental biology*, ed. A. A. Moscona and A. Monroy, vol. 12, pp. 186–246. New York: Academic Press.

Eppig, J. J. (1978) A comparison between oocyte growth in coculture with granulosa cells and oocytes with granulosa cell-oocyte junctional contact maintained in vitro. *J. exp. Zool. 209*, 345–53.

Epstein, C. J. (1969) Mammalian oocytes: X chromosome activity. *Science 163*, 1078–79.

———. (1972) Expression of the mammalian X chromosome before and after fertilisation. *Science 175*, 1467–68.

Epstein, C. J., Smith, S., Travis, B., and Tucker, G. (1978) Both X chromosomes function before visible X-chromosome inactivation in female mouse embryos. *Nature 274*, 500–03.

Erickson, J. D. (1978) Down syndrome, paternal age, maternal age and birth order. *Ann. hum. Genet. 41*, 289–98.

Evans, E. P., Ford, C. E., and Lyon, M. F. (1977) Direct evidence of the capacity of the XY germ cell in the mouse to become an oocyte. *Nature 267*, 430–31.

Faddy, M. J., Jones, E. C., and Edwards, R. G. (1976) An analytical model for ovarian follicle dynamics. *J. exp. Zool. 197*, 173–85.

Fawcett, D. W., Eddy, E. M., and Phillips, D. M. (1970) Observations on the fine structure and relationships of the chromatoid body in mammalian spermatogenesis. *Biol. Reprod.* 2, 129–53.

Ford, C. E. (1960) Chromosomal abnormality and congenital malformation. In *Congenital malformations*. Ciba Foundation Symposium, ed. G. E. W. Wolstenholme and C. M. O'Connor, pp. 32–47. London and Edinburgh: Churchill Livingstone.

Ford, C. E., and Evans, E. P. (1977) Cytogenetic observations on XX/XY chimaeras and a reassessment of the evidence for germ cell chimaerism in heterosexual twin cattle and marmosets. *J. Reprod. Fert. 49*, 25–33.

Ford, C. E., Evans, E. P., Burtenshaw, M. D., Clegg, H. M., Tuffrey, M., and Barnes, R. D. (1975) A functional "sex-reversed" oocyte in the mouse. *Proc. Roy. Soc. Lond. B*, *190*, 187–97.

Frels, W. I., and Chapman, V. M. (1979) Paternal X chromosome expression in extraembryonic membranes of XO mice. *J. exp. Zool. 210*, 553–60.

Fulton, B. F., and Whittingham, D. G. (1978) Activation of mammalian oocytes by intracellular injection of calcium. *Nature 273*, 149–51.

Gabel, C. Q., Eddy, E. M., and Shapiro, B. M. (1979) After fertilization, sperm surface components remain as a patch in sea urchin and mouse embryos. *Cell 18*, 207–16.

Gardner, R. L. (1977) Developmental potency of normal and neoplastic cells of the early mouse embryo. In *Birth defects*. Excerpta Medica International Congress Ser. 432, ed. J. W. Littlefield and J. de Grouchy, pp. 154–66. Amsterdam and Oxford: Excerpta Medica.

Gardner, R. L., and Johnson, M. H. (1972) An investigation of inner cell mass and trophectoderm tissues following their isolation from the mouse blastocyst. *J. Embryol. exp. Morph. 28*, 279–312.

Gardner, R. L., and Lyon, M. F. (1971) X chromosome inactivation studied by injection of a single cell into the mouse blastocyst. *Nature 231*, 385–86.

Gardner, R. L., Papaioannou, V. E., and Barton, S. C. (1973) Origin of the ectoplacental cone and secondary giant cells in mouse blastocysts reconstituted from isolated trophoblast and inner cell mass. *J. Embryol. exp. Morph. 30*, 561–72.

Gardner, R. L., and Rossant, J. (1976) Determination during embryogenesis. In *Embryogenesis in mammals*, Ciba Foundation Symposium, vol. 40 (new series), ed. K. Elliott and M. O'Connor, pp. 5–25. Amsterdam: Elsevier.

Gardner, R. L., and Rossant, J. (1979) Investigation of the fate of 4.5 day *post-coitum* mouse inner cell mass cells by blastocyst injection. *J. Embryol. exp. Morph. 52*, 141–52.

Gartler, S. M., Andina, R., and Gant, N. (1975) Ontogeny of X-chromosome inactivation in the female germ line. *Exp. Cell Res. 91*, 454–57.

Gartler, S. M., Liskay, R. M., and Gant, N. (1973) Two functional X-chromosomes in human fetal oocytes. *Exp. Cell Res. 82*, 464–66.

German, J. (1968) Mongolism, delayed fertilization and human sexual behaviour. *Nature 217*, 516–18.

Ghosal, S. K., and Mukherjee, B. B. (1977) Replicative differentiation of Y chromosome in mammalian testis. *Nucleus 20*, 55–60.

Gilula, N. B., Epstein, M. L., and Beers, W. H. (1978) Cell-to-cell communication and ovulation. A study of the cumulus-oocyte complex. *J. Cell Biol. 78*, 58–75.

Glass, L. E. (1961) Localization of autologous and heterologous serum antigens in the mouse ovary. *Dev. Biol. 3*, 787–804.

Glass, L. E., and Cons, J. M. (1968) Stage dependent transfer of systemically injected foreign protein, antigen and radiolabel into mouse ovarian follicles. *Anat. Rec. 162*, 139–56.

Graham, C. F. (1971) The design of the mouse blastocyst. In *Control mechanisms of growth and differentiation*, ed. D. Davis and M. Balls, Symp. Soc. exp. Biol., vol. 25, 371–78. Cambridge: Cambridge University Press.

———. (1974) The production of parthenogenetic mammalian embryos and their use in biological research. *Biol. Rev. 49*, 399–422.

Graham, C. F., and Deussen, Z. A. (1978) Features of cell lineages in preimplantation mouse development. *J. Embryol. exp. Morph. 48*, 53–72.

Grinsted, J., Byskov, A. G., and Andreasen, M. P. (1979) Induction of meiosis in fetal mouse testis *in vitro* by rete testis tissue from pubertal mice and bulls. *J. Reprod. Fert. 56*, 653–56.

Haddad, A., and Nagai, M. E. T. (1977) Radioautographic study of glycoprotein biosynthesis and renewal on the ovarian follicles of mice and the origin of the zona pellucida. *Cell Tiss. Res. 177*, 347–69.

Handyside, A. H. (1978) Time of commitment of inside cells isolated from preimplantation mouse embryos. *J. Embryol. exp. Morph. 45*, 37–53.

Handyside, A. H., and Barton, S. C. (1977) Evaluation of the technique of immunosurgery for the isolation of inner cell masses from mouse blastocysts. *J. Embryol. exp. Morph. 37*, 217–26.

Handyside, A. H., and Johnson, M. H. (1978) Temporal and spatial patterns of synthesis of tissue-specific polypeptides in the pre-implantation mouse embryo. *J. Embryol. exp. Morph. 44*, 191–99.

Hassold, T., Jacobs, P., Kline, J., Stein, Z., and Warburton, D. (1980) Effect of maternal age on autosomal trisomies. *Ann. hum. Genet. 44*, 29–36.

Heasman, J., Mohun, T., and Wylie, C. C. (1977) Studies on the locomotion of primordial germ cells from *Xenopus laevis in vitro*. *J. Embryol. exp. Morph. 42*, 149–61.

Heasman, J., and Wylie, C. C. (1978) Electron microscopic studies on the structure of motile primordial germ cells of *Xenopus laevis in vitro*. *J. Embryol. exp. Morph. 46*, 119–33.

Henderson, S. A., and Edwards, R. G. (1968) Chiasma frequency and maternal age in mammals. *Nature* (London) *218*, 22–28.

Hillman, N., Sherman, M. I., and Graham, C. (1972) The effect of spatial arrangement on cell determination during mouse development. *J. Embryol. exp. Morph. 28*, 263–78.

Hilscher, W., and Hilscher, B. (1976) Kinetics of the male gametogenesis. *Andrologia 8*, 105–16.

Hogan, B., and Tilly, R. (1978) *In vitro* development of inner cell masses isolated immunosurgically from mouse blastocysts. I. Inner cell masses from 3.5-day p.c. blastocysts incubated for 24 h before immunosurgery. II. Inner cell masses from 3.5- to 4.0-day p.c. blastocysts. *J. Embryol. exp. Morph. 45*, 93–105, 107–21.

Hoppe, P. C., and Illmensee, K. (1977) Microsurgically produced homozygous-diploid uniparental mice. *Proc. Nat. Acad. Sci. U.S.A. 74*, 5657–61.

Howe, C. C., Gmür, R., and Solter, D. (1980) Cytoplasmic and nuclear protein synthesis during *in vitro* differentiation of murine ICM and embryonal carcinoma cells. *Dev. Biol. 74*, 351–63.

Illmensee, K., & Hoppe, P. C. (1981). Nuclear transplantation in Mus musculus: Developmental potential of nuclei from preimplantation embryos. *Cell 23*, 9–18.

Illmensee, K., and Mahowald, A. P. (1974) Transplantation of posterior pole plasm in Drosophila. Induction of germ cells at the anterior pole of the egg. *Proc. Nat. Acad. Sci. U.S.A. 71*, 1016–20.

Iwamatsu, T., and Yanagimachi, R. (1975) Maturation *in vitro* of ovarian oocytes of prepubertal and adult hamsters. *J. Reprod. Fert. 45*, 83–90.

Johnson, M. H. (1979) Molecular differentiation of inside cells and inner cell masses isolated from the preimplantation mouse embryo. *J. Embryol. exp. Morph. 53*, 335–44.

Johnson, M. H., Chakraborty, J., Handyside, A. H., Willison, K., and Stern, P. (1979) The effect of prolonged decompaction on the development of the preimplantation mouse embryo. *J. Embryol. exp. Morph. 54*, 241–61.

Jones, E. C., and Krohn, P. L. (1961) The effect of hypophysectomy on age changes in the ovaries of mice. *J. Endocrin. 21*, 497–509.

Kaufman, M. H., Barton, S. C., and Surani, M. A. H. (1977) Normal post-implantation development of mouse parthenogenetic embryos to the forelimb bud stage. *Nature* (London) *265*, 53–55.

Kaufman, M. H., Guc-Cubrilo, M., and Lyon, M. F. (1978) X chromosome inactivation in diploid parthenogenetic mouse embryos. *Nature 271*, 547–49.

Kellokumpu-Lehtinen, P., and Söderström, K. (1978) Occurrence of nuage in fetal human germ cells. *Cell Tiss. Res. 194*, 171–77.

Kelly, S. J. (1977) Studies of the developmental potential of 4- and 8-cell stage mouse blastomeres. *J. exp. Zool. 200*, 365–76.

Kelly, S. J., Mulnard, J. G., and Graham, C. F. (1978) Cell division and cell allocation in early mouse development. *J. Embryol. exp. Morph. 48*, 37–51.

Kindred, B. M. (1961) A maternal effect on vibrissae score due to the *Tabby* gene. *Austr. J. Biol. Sci. 14*, 627–36.

Kozak, L. P., and Quinn, P. J. (1975) Evidence for dosage compensation of an X-linked gene in the 6-day embryo of the mouse. *Dev. Biol. 45*, 65–73.

Krarup, T., Pedersen, T., and Faber, M. (1969) Regulation of oocyte growth in the mouse ovary. *Nature 224*, 187–88.

Kratzer, P. G., and Gartler, S. M. (1978) HPRT activity changes in preimplantation mouse embryos. *Nature 274*, 503–04.

LaMarca, M. J., and Wassarman, P. M. (1979) Program of early development in the mammal: changes in absolute rates of synthesis of ribosomal proteins during oogenesis and early embryogenesis in the mouse. *Dev. Biol. 73*, 103–19.

Lansing, A. I. (1952) Biological and cellular problems of ageing: 1. General physiology. In *Cowdry's problem of ageing*. 3rd ed., pp. 14–19. Baltimore: Williams and Wilkins.

Lee, H., Karasanyi, N., and Nagele, R. G. (1978) The role of the cell surface in the migration of primordial germ cells in early chick embryos: effects of concanavalin A. *J. Embryol. exp. Morph. 46*, 5–20.

Lifschytz, E., and Lindsley, D. L. (1972) The role of X-chromosome inactivation during spermatogenesis. *Proc. Nat. Acad. Sci. U.S.A. 69*, 182–86.

Lo, C. W., and Gilula, N. B. (1980) Gap junctional communication in the preimplantation mouse embryo. *Cell 18*, 399–409.

Lyon, M. F. (1972) X chromosome inactivation and developmental patterns in mammals. *Biol. Rev. 47*, 1–35.

———. (1974) Sex chromosome activity in germ cells. In *Physiology and genetics of reproduction A*, ed. E. M. Coutinho and F. Fuchs, pp. 63–71. New York and London: Plenum.

Lyon, M. F., Glenister, P. H., and Lamoreux, M. L. (1975). Normal spermatozoa from androgen-resistant germ cells of chimaeric mice and the role of androgen in spermatogenesis. *Nature* (London) *258*, 620–22.

McCarrey, J. R., and Abbott, U. K. (1978) Chick gonad differentiation following excision of primordial germ cells. *Dev. Biol. 66*, 256–65.

McCoshen, J. A., and McCallion, D. J. (1975) A study of the primordial germ cells during their migratory phase in Steel mutant mice. *Experientia 31*, 589–90.

McCulloch, E. A., Siminovitch, L., Till, J. E., Russell, E. S., and

Bernstein, S. E. (1965) The cellular basis of the genetically determined hemopoietic defect in anemic mice of genotype *Sl/Sl^d^*. *Blood 26*, 399–410.

McCuskey, R. S., and Meineke, H. A. (1973) Studies of the hemopoietic microenvironment. III. Differences in the splenic microvascular system and stroma between Sl/Sl^d and W/W^v anemic mice. *Amer. J. Anat. 137*, 187–98.

McLaren, A. (1975a) The independence of germ-cell genotype from somatic influence in chimaeric mice. *Genet. Res. 25*, 83–87.

———. (1975b) Sex chimaerism and germ cell distribution in a series of chimaeric mice. *J. Embryol. exp. Morph. 33*, 205–16.

———. (1976a) Genetics of the early mouse embryo. *A. Rev. Genet. 10*, 361–88.

———. (1976b) Growth from fertilization to birth in the mouse. In *Embryogenesis in mammals*, Ciba Foundation Symposium, vol. 40 (new series), ed. K. Elliott and M. O'Connor, pp. 47–51. Amsterdam: Elsevier.

———. (1976c) *Mammalian chimaeras*. Cambridge: Cambridge University Press.

———. (1979) The impact of pre-fertilization events on post-fertilization development in mammals. In *Maternal effects in development*, ed. D. R. Newth and M. Balls, Brit. Soc. Dev. Biol. Symp., vol. 4, pp. 287–320. Cambridge: Cambridge University Press.

———. (1980) Oocytes in the testis. *Nature 283*, 688–89.

———. (1981) The fate of germ cells in the testis of fetal *Sex-reversed* mice. *J. Reprod. Fert. 61*, 461–67.

McLaren, A., Chandley, A. C., and Kofman-Alfaro, S. (1972) A study of meiotic germ cells in the gonads of foetal mouse chimaeras. *J. Embryol. exp. Morph. 27*, 515–24.

McLaren, A., and Monk, M. (1981) X chromosome activity in the germ cells of Sex-reversed mouse embryos. *J. Reprod. Fert.*, in press.

Markert, C. L., and Petters, R. M. (1977) Homozygous mouse embryos produced by microsurgery. *J. exp. Zool. 201*, 295–301.

Martin, G. R., Epstein, C. J., Travis, B., Tucker, G., Yatziv, S., Martin, D. W., Clift, S., and Cohen, S. (1978) X-chromosome inactivation during differentiation of female teratocarcinoma stem cells *in vitro*. *Nature 271*, 329–33.

Masui, Y., and Pedersen, R. A. (1975) Ultraviolet-light induced unscheduled DNA synthesis in mouse oocytes during meiotic maturation. *Nature 257*, 705–06.

Mayer, T. C. (1973) Site of gene action in steel mice. Analysis of the pigment defect by mesoderm-ectoderm recombinations. *J. exp. Zool. 184*, 345–52.

Mayer, T. C., and Green, M. C. (1968) An experimental analysis of the pigment defect caused by mutations at the *W* and *Sl* loci in mice. *Dev. Biol 18*, 62–75.

Merchant, H. (1975) Rat gonadal and ovarian organogenesis with and without germ cells. An ultrastructural study. *Dev. Biol. 44*, 1–21.

Merchant-Larios, H., and Coello, J. (1979) The effect of busulfan on rat primordial germ cells at the ultrastructural level. *Cell Differentiation 8*, 145–55.

Meyerhof, P. G., and Masui, Y. (1977) Ca and Mg control of cytostatic factors from Rana pipiens which cause metaphase and cleavage arrest. *Dev. Biol. 61*, 214–29.

Michie, D. (1958) The third stage in genetics. In *A century of Darwin*, ed. S. A. Barnett. London: Heinemann.

Migeon, B. R., and Jelalian, K. (1977) Evidence for two active X chromosomes in germ cells of female before meiotic entry. *Nature 269*, 242–43.

Mintz, B., and Russell, E. S. (1957) Gene-induced embryological modifications of primordial germ cells in the mouse. *J. exp. Zool. 134*, 207–38.

Mittwoch, U., and Buehr, M. L. (1973) Gonadal growth in embryos of sex reversed mice. *Differentiation. 1*, 219–24.

Modlinski, J. A. (1975) Haploid mouse embryos obtained by microsurgical removal of one pronucleus. *J. Embryol. exp. Morph. 33*, 897–905.

Monesi, V. (1971) Chromosome activities during meiosis and spermiogenesis. *J. Reprod. Fert.*, Suppl. *13*, 1–9.

Monk, M. (1978) Biochemical studies on mammalian X-chromosome activity. In *Development in mammals*, ed. M. H. Johnson, vol. *3*, pp. 189–223. Amsterdam: Elsevier.

Monk, M., and Harper, M. (1978) X-chromosome activity in preimplantation embryos from XX and XO mothers. *J. Embryol. exp. Morph. 46*, 53–64.

Monk, M., and Harper, M. I. (1979) Sequential X-chromosome inactivation coupled with cellular differentiation in early mouse embryos. *Nature 281*, 311–13.

Monk, M., and McLaren, A. (1981) X-chromosome activity in fetal germ cells of the mouse. *J. Embryol. exp. Morph.*, in press.

Moor, R. M. (1978) Role of steroids in the maturation of ovarian oocytes. *Annls Biol. anim. Biochim. Biophys. 18*, 477–82.

Moor, R. M., and Cran, D. G. (1980) Intercellular coupling in mammalian oocytes. In *Development in mammals*, ed. M. H. Johnson, vol. 4, pp. 3–37. Amsterdam: Elsevier.

Moor, R. M., Polge, C., and Willadsen, S. M. (1980) Effect of follicular steroids on the maturation and fertilization of mammalian oocytes. *J. Embryol. exp. Morph. 56*, 319–35.

Moor, R. M., Smith, M. W., and Dawson, R. M. C. (1980) Measurement of intercellular coupling between oocytes and cumulus cells using intracellular markers. *Exp. Cell Res., 126*, 15–29.

Moor, R. M., and Trounson, A. O. (1977) Hormonal and follicular factors affecting maturation of sheep oocytes *in vitro* and their subsequent developmental capacity. *J. Reprod. Fert. 49*, 101–09.

Moor, R. M., and Warnes, G. M. (1978) Regulation of oocyte maturation in mammals. In *Control of ovulation*, ed. D. B. Crighton, G. R. Foxcroft, N. B. Haynes, and G. E. Lamming. London: Butterworths.

Mukherjee, A. B. (1972) Normal progeny from fertilization *in vitro* of mouse oocytes matured in culture and spermatozoa capacitated *in vitro*. *Nature 237*, 397–98.

Mystkowska, E. T., and Tarkowski, A. K. (1968) Observations on CBA-p/CBA-T6T6 mouse chimeras. *J. Embryol. exp. Morph. 20*, 33–52.

Mystkowska, E. T., and Tarkowski, A. K. (1970) Behaviour of germ cells and sexual differentiation in late embryonic and early postnatal mouse chimaeras. *J. Embryol. exp. Morph. 23*, 395–405.

Nicosia, S. V., Wolf, D. P., and Inoue, M. (1977) Cortical granule distribution and cell surface characteristics in mouse eggs. *Dev. Biol. 57*, 56–74.

Noda, Y. D., and Yanagimachi, R. (1976) Electron microscopic observations of guinea pig spermatozoa penetrating eggs *in vitro*. *Dev. Growth Diff. 18*, 15–23.

Nussbaum, M. (1880) Die Differenzierung des Geschlechts in Thierreich. *Arch. Mikrosk. Anat. Entwicklungsmech. 18*, 1–121.

O, W. S., and Baker, T. G. (1976) Initiation and control of meiosis in hamster gonads *in vitro*. *J. Reprod. Fert. 48*, 399–401.

O, W. S., and Baker, T. G. (1978) Germinal and somatic cell interrelationships in gonadal sex differentiation. *Annls Biol. anim. Biochim. Biophys. 18*, 351–57.

Ohno, S. (1964) Life history of the female germ cells in mammals. In *Proceedings of the Second International Conference on Congenital Malformations*, pp. 36–50. New York: National Foundation.

Ohno, S., and Gropp, A. (1965) Embryological basis for germ cell chimerism in mammals. *Cytogenetics 4*, 251–61.

Ohno, S., Kaplan, W. D., and Kinosita, R. (1961) X-chromosome behaviour in germ and somatic cells of Rattus norvegicus. *Exp. Cell Res. 22*, 535–44.

Ohno, S., Nagai, Y., and Ciccarese, S. (1978) Testicular cells lysostripped of H-Y antigen organize ovarian follicle-like aggregates. *Cytogenet. Cell Genet. 20*, 351–64.

Ohno, S., Trujillo, J. M., Stenius, C., Christian, L. C., and Teplitz, R. L. (1962) Possible germ cell chimeras among newborn dizygotic twin calves (*Bos Taurus*). *Cytogenetics 1*, 258–65.

Ożdżeński, W. (1967) Observations on the origin of primordial germ cells in the mouse. *Zoologica Pol. 17*, 367–79.

———. (1972) Differentiation of the genital ridges of mouse embryos in the kidney of adult mice. *Arch. Anat. Micr. Morph. Exp. 61*, 267–78.

Ożdżeński, W., Rogulska, T., Balakier, H., Brzozowska, M., Rembiszewska, A., and Stepinska, U. (1976) Influence of embryonic and adult testis on the differentiation of embryonic ovary in the mouse. *Arch. Anat. Microscop. 65*, 285–94.

Parkening, T. A., and Chang, M. C. (1976) In vitro fertilization of ova from senescent mice and hamsters. *J. Reprod. Fert. 48*, 381–83.

Rodman, T. C., and Bachvarova, R. (1976) RNA synthesis in preovulatory mouse oocytes. *J. Cell Biol. 70*, 251–57.

Rogulska, T., Ożdżeński, W., and Komar, A. (1971) Behaviour of mouse primordial germ cells in the chick embryo. *J. Embryol. exp. Morph. 85*, 155–64.

Rossant, J. (1975) Investigation of the determinative state of the mouse inner cell mass. I. Aggregation of isolated inner cell masses with morulae. *J. Embryol. exp. Morph. 33*, 979–90.

Rossant, J., Gardner, R. L., and Alexandre, H. L. (1978) Investigation of the potency of cells from the postimplantation mouse embryo by blastocyst injection: a preliminary report. *J. Embryol. exp. Morph. 48*, 239–47.

Rossant, J., and Vijh, K. M. (1980) Ability of outside cells from preimplantation mouse embryos to form inner cell mass derivatives. *Dev. Biol. 76*, 475–82.

Schultz, R. M., Letourneau, G. E., and Wassarman, P. M. (1979) Program of early development in the mammal: Changes in the pattern and absolute rates of tubulin and total protein synthesis during oocyte growth in the mouse. *Dev. Biol. 73*, 120–33.

Semyonova-Tian-Shanskaya, A. G. (1976) Cytological features of gonocytes in the course of their migration in early embryos of rats. *Arkh. Anat. Gistol. Embriol. 71*, 52–58. (in Russian)

Semyonova-Tian-Shanskaya, A. G., and Patkin, E. L. (1978) Changes in gonocyte nuclei at different stages of their differentiation in early human female embryos. *Arkh. Anat. Gistol. Embriol. 74*, 91–97. (in Russian)

Sherman, M. I. (1979) Developmental biochemistry of preimplantation mammalian embryos. *A. Rev. Biochem. 48*, 443–70.

Skreb, N., Svajger, A., and Levak-Svajger, B. (1976) Developmental potentialities of the germ layers in mammals. In *Embryogenesis in mammals*, Ciba Foundation Symposium 40 (new series), ed. K. Elliott and M. O'Connor, pp. 27–39. Amsterdam: Associated Scientific Publishers.

Smith, L. D., and Williams, M. (1979) Germinal plasm and germ cell determinants in anuran amphibians. In *Maternal effects in development*. Brit. Soc. Dev. Biol. Symp., vol. 4, ed. D. R. Newth and M. Balls, pp. 167–97. Cambridge: Cambridge University Press.

Smith, R., and McLaren, A. (1977) Factors affecting the time of formation of the mouse blastocoele. *J. Embryol. exp. Morph. 41*, 79–92.

Snow, M. H. L. (1977) Gastrulation in the mouse: Growth and regionalization of the epiblast. *J. Embryol. exp. Morph. 42*, 293–303.

Snow, M. H. L., and Bennett, D. (1978) Gastrulation in the mouse: assessment of cell populations in the epiblast of t^{w18}/t^{w18} embryos. *J. Embryol. exp. Morph. 47*, 39–52.

Snow, M. H. L., and Tam, P. P. L. (1979) Is compensatory growth a complicating factor in mouse teratology? *Nature 279*, 555–57.

Spiegelman, M., and Bennett, D. (1973) A light- and electron-microscopic study of primordial germ cells in the early mouse embryo. *J. Embryol. exp. Morph. 30*, 97–118.

Spindle, A. I. (1978) Trophoblast regeneration by inner cell masses isolated from cultured mouse embryos. *J. exp. Zool. 203*, 483–89.

Stevens, L. C. (1978) Totipotent cells of parthenogenetic origin in chimaeric mice. *Nature 176*, 266–67.

Stevens, L. C., and Varnum, D. S. (1974) The development of teratomas from parthenogenetically activated ovarian mouse eggs. *Dev. Biol. 37*, 369–80.

Stevens, L. C., Varnum, D. S., and Eicher, E. M. (1977) Viable chimaeras produced from normal and parthenogenetic mouse embryos. *Nature* (London) *269*, 515–17.

Surani, M. A. H., Barton, S. C., and Kaufman, M. H. (1977) Development to term of chimaeras between diploid parthenogenetic and fertilized embryos. *Nature* (London) *270*, 601–03.

Szöllösi, D., Gérard, M., Ménézo, Y., and Thibault, C. (1978) Permeability of ovarian follicle; corona cell-oocyte relationship in mammals. *Annls Biol. anim. Biochim. Biophys. 18*, 511–21.

Takagi, N., and Sasaki, M. (1975) Preferential inactivation of the paternally derived X chromosome in extraembryonic membranes of the mouse. *Nature 256*, 640–42.

Takagi, N., Wake, N., and Sasaki, M. (1978) Cytologic evidence for preferential inactivation of the paternally derived X chromosome in XX mouse blastocysts. *Cytogenet. Cell Genet. 20*, 240–48.

Tam, P., and Snow, M. H. L. (1981) Proliferation and migration of primordial germ cells during compensatory growth in the mouse embryo. *J. Embryol. exp. Morph.*, in press.

Tarkowski, A. K. (1959) Experiments on the development of isolated blastomeres of mouse eggs. *Nature 184*, 1286–87.

Thibault, C. (1972) Final stages of mammalian oocyte maturation. In *Oogenesis*, ed. J. D. Biggers and A. W. Schultz, pp. 397–411. Baltimore: University Park Press.

———. (1977) Are follicular maturation and oocyte maturation independent processes? *J. Reprod. Fert. 51*, 1–15.

Turner, C. D. (1966) *General Endocrinology*. 5th ed. Philadelphia: W. B. Saunders.

———. (1969) Experimental reversal of germ cells. *Embryologia 10*, 206–30.

Van Blerkom, J. (1979) Molecular differentiation of the rabbit ovum. III. Fertilization-autonomous polypeptide synthesis. *Dev. Biol. 72*, 188–94.

Van Blerkom, J., and Brockway, G. O. (1975) Qualitative patterns of protein synthesis in the preimplantation mouse embryo. I. Normal pregnancy. *Dev. Biol. 44*, 148–57.

Van Blerkom, J., and McGaughey, R. W. (1978) Molecular differentiation of the rabbit ovum. I. During oocyte maturation *in vivo* and *in vitro*. *Dev. Biol. 63*, 151–64.

Verrusio, A. C., Pollard, D. R., and Fraser, F. C. (1968) A cytoplasmically transmitted, diet-dependent difference in response to the teratogenic effects of 6-aminonicotinamide. *Science 160*, 206–07.

Wachtel, S. S., and Koo, G. C. (1981) H-Y antigen in gonadal differentiation. In *Mechanisms of sex differentiation in animals and man*, ed. C. R. Austin and R. G. Edwards. London: Academic Press. In press.

Wachtel, S. S., Ohno, S., Koo, G. C., and Boyse, E. A. (1975) Possible role for H-Y antigen in the primary determination of sex. *Nature 257*, 235–36.

Wakasugi, N. (1973) Studies on fertility of DDK mice: Reciprocal crosses between DDK and C57BL/6J strains and experimental transplantation of the ovary. *J. Reprod. Fert. 33*, 283–91.

———. (1974) A genetically determined incompatibility system between spermatozoa and eggs leading to embryonic death in mice. *J. Reprod. Fert. 41*, 85–96.

Wake, N., Takagi, N., and Sasaki, M. (1976) Non-random inactivation of X-chromosome in the rat yolk sac. *Nature 262*, 580–87.

Warnes, G. M., Moor, R. M., and Johnson, M. H. (1977) Changes in protein synthesis during maturation of sheep oocytes *in vivo* and *in vitro*. *J. Reprod. Fert. 49*, 331–35.

Weismann, A. (1892) Das Keimplasma. Eine theorie der Vererburg. Jena: Gustav Fischer.

West, J. D., Frels, W. I., Chapman, V. M., and Papaioannou, V. E. (1977) Preferential expression of the maternally derived X chromosomes in the mouse yolk sac. *Cell 12*, 873–82.

West, J. D., Papaioannou, V. E., Frels, W. I., and Chapman, V. M. (1978) Preferential expression of the maternally derived X chromosome in extraembryonic tissues of the mouse. In *Genetic mosaics and chimeras in mammals*, ed. L. B. Russell, pp. 361–77. New York: Plenum.

Whittingham, D. G. (1980) *Parthenogenesis in mammals*. Oxford Reviews of Reproductive Biology, vol. 2, 205–31. Oxford: Oxford University Press.

Whittingham, D. G., and Siracusa, G. (1978) The involvement of calcium in the activation of mammalian oocytes. *Exp. Cell Res. 113*, 311–17.

Willadsen, S. M. (1978) A method for culture of micromanipulated sheep embryos and its use to produce monozygotic twins. *Nature 277*, 298–300.

Willadsen, S. M., and Polge, C. (1981) Attempts to produce monozygotic quadruplets in cattle. *Vet. Rec. 108*, 211–13.

Witschi, E. (1948) Migration of the germ cells of human embryos from the yolk sac to the primitive gonadal folds. *Carnegie Inst. Contrib. Embryol. 32*, 67–80.

Wolff, G. L. (1978) Influence of maternal phenotype on metabolic differentiation of agouti locus mutants in the mouse. *Genetics 88*, 529–39.

Wylie, C. C., Heasman, J., Swan, A. P., and Anderton, B. H. (1979) Evidence for substrate guidance of primordial germ cells. *Exp. Cell Res. 121*, 315–24.

Zenzes, M. T., Wolf, V., Gunther, E., and Engel, W. (1978) Studies on the function of H-Y antigen: dissociation and reorganization experiments on rat gonadal tissue. *Cytogenet. Cell Genet. 20*, 365–72.

Ziomek, C. A., and Johnson, M. H. (1981) Cell surface interaction induces polarization of mouse 8-cell blastomeres at compaction. *Cell 21*, 935–42.

Index

LADY MARGARET HALL LIBRARY
OXFORD

SILLIMAN VOLUMES IN PRINT

Jacob Bronowski, *The Origins of Knowledge and Imagination*
S. Chandrasekhar, *Ellipsoidal Figures of Equilibrium*
Theodosius Dobzhansky, *Mankind Evolving*
René Dubos, *Man Adapting*
Ross Granville Harrison, *Organization and Development of the Embryo*
John von Neumann, *The Computer and the Brain*
Karl K. Turekian, *Late Cenozoic Glacial Ages*

LADY MARGARET HALL LIBRARY
OXFORD